The Jemez Mountains

Frontispiece. Young girls and hot spring gazebo. Photo by R. B. Townshend, 1903. (Courtesy Pitt Rivers Museum, University of Oxford, Catalog No. 1998.58.64)

THE
Jemez Mountains

A Cultural and
Natural History

Thomas W. Swetnam

UNIVERSITY OF NEW MEXICO PRESS ——— ALBUQUERQUE

© 2025 by Thomas W. Swetnam
All rights reserved. Published 2025
Printed in the United States of America

Library of Congress Cataloging-in-Publication Data
Names: Swetnam, Thomas W., author.
Title: The Jemez Mountains: a cultural and natural history / Thomas W. Swetnam.
Description: Albuquerque: University of New Mexico Press, 2025. | Includes bibliographical references and index.
Identifiers: LCCN 2024032531 (print) | LCCN 2024032532 (ebook) | ISBN 9780826367754 (paperback) | ISBN 9780826367761 (epub)
Subjects: LCSH: Jemez Mountains (N.M.)—History. | Jemez Mountains (N.M.)—Social life and customs. | Jemez Mountains (N.M.)—Social conditions.
Classification: LCC F802.S3 S94 2025 (print) | LCC F802.S3 (ebook) | DDC 978.957--dc23/eng/20240716
LC record available at https://lccn.loc.gov/2024032531
LC ebook record available at https://lccn.loc.gov/2024032532

Founded in 1889, the University of New Mexico sits on the traditional homelands of the Pueblo of Sandia. The original peoples of New Mexico—Pueblo, Navajo, and Apache—since time immemorial have deep connections to the land and have made significant contributions to the broader community statewide. We honor the land itself and those who remain stewards of this land throughout the generations and also acknowledge our committed relationship to Indigenous peoples. We gratefully recognize our history.

Cover photograph by James McCue
Designed by Isaac Morris
Composed in Adobe Jenson, Canto, and Condo

*To the memory of my parents, Fred and Grace,
and my brother and sister-in-law, Mike and Karen*

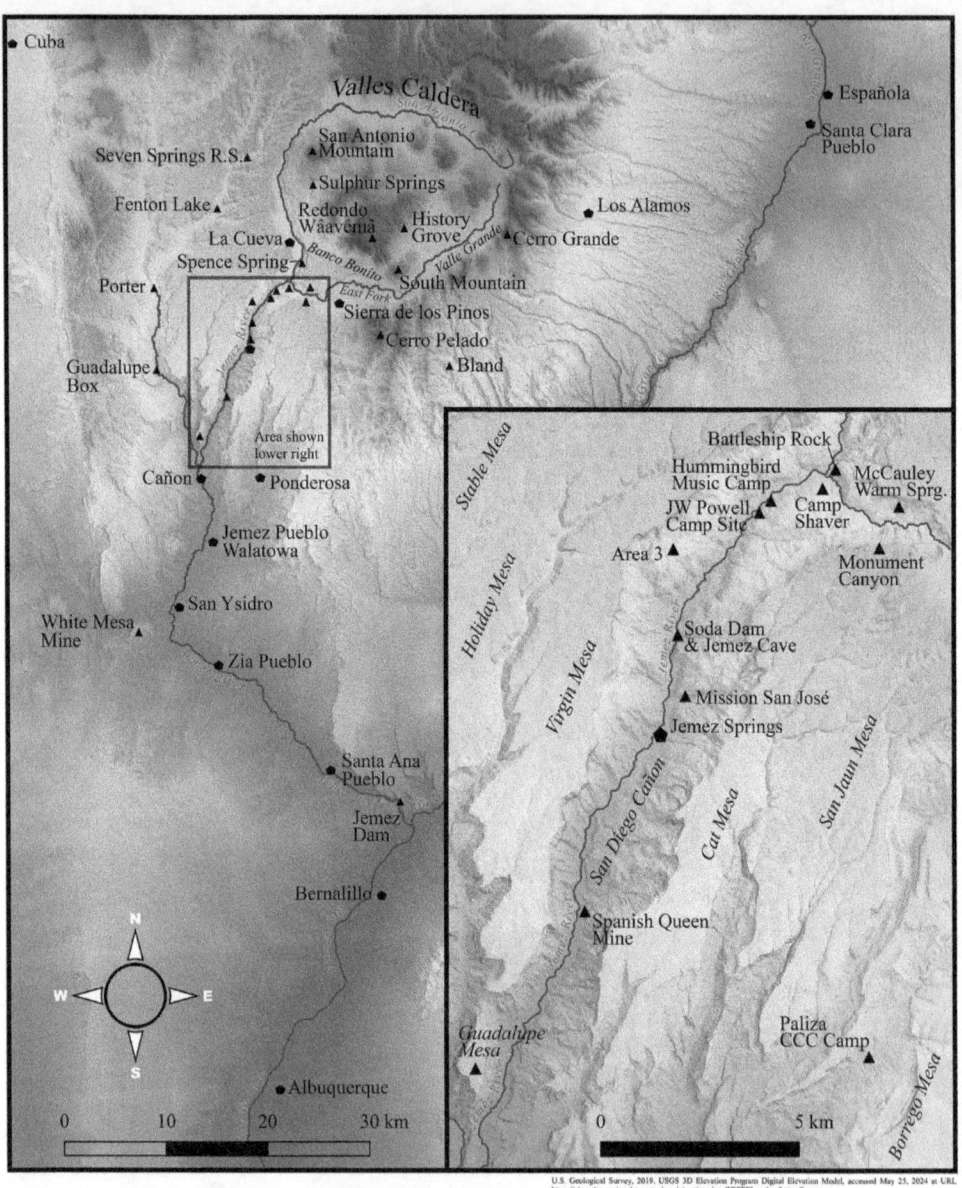

CONTENTS

PREFACE —— xi

ACKNOWLEDGMENTS —— xiv

Part I — Stories from Past Centuries

CHAPTER 1
The Hemish Footprint and Spanish Entrada —— 3

CHAPTER 2
San José de los Jémez Mission —— 9

CHAPTER 3
Pueblo Revolts and Battles of Reconquest —— 17

CHAPTER 4
Treasures of the Spanish Queen Mine —— 27

CHAPTER 5
Hillers and the Stevensons in the Jemez —— 33

CHAPTER 6
Myths and Legends of Montezuma in New Mexico —— 39

CHAPTER 7
A Visit to the Valles in 1886 —— 45

CHAPTER 8
John Wesley Powell's Jemez Dam Dreams —— 53

CHAPTER 9
Dry Goods, Saloons, and Sheep —— 59

CHAPTER 10
First Automobile in Jemez Springs —— 66

CHAPTER 11
Whose Lands Are the Jemez Mountains? —— 71

CHAPTER 12
The Santa Fe Northwestern Railroad —— 77

Contents

CHAPTER 13
The Jemez National Forest —— 85

CHAPTER 14
Peeled Ponderosa Pines —— 95

CHAPTER 15
Four Jemez Mountains Grizzly Bear Stories —— 101

Part II   Stories from Richard Baxter Townshend

CHAPTER 16
Wild West Days of the Jemez Valley —— 109

CHAPTER 17
Penitentes in the Jemez Valley —— 115

CHAPTER 18
Correr El Gallo at Walatowa —— 121

CHAPTER 19
Presbyterians in the Jemez —— 127

CHAPTER 20
A Visit to the Rio Cebolla in 1903 —— 137

CHAPTER 21
Horse Logging Above Jemez Springs —— 145

Part III   Soda Dam, Geology, and Floods

CHAPTER 22
Soda Dam, Logs, and Floods —— 155

CHAPTER 23
The Many Soda Dams and Lakes of the Valles Caldera —— 161

CHAPTER 24
The Blasting of Soda Dam —— 167

CHAPTER 25
The Caves of Soda Dam —— 174

CHAPTER 26
The Explosion Craters of Banco Bonito —— 179

CHAPTER 27
Great Floods of the Rio Jemez —— 183

CHAPTER 28
Fossils in the Jemez —— 191

Part IV Fire, Forests, and Cottonwoods

CHAPTER 29
Forests and People on the Southern Jemez Plateau —— 199

CHAPTER 30
The Era of Runaway Wildfires —— 205

CHAPTER 31
Bald Mountains in the Jemez —— 211

CHAPTER 32
Cerro Pelado Fire and Using All the Tools —— 217

CHAPTER 33
Cottonwoods and Junipers in the Valley —— 223

CHAPTER 34
Climate Change in the Jemez —— 231

Part V Hotels, Hot Springs, Hippies, Campers, and Priests

CHAPTER 35
Hotels, Mines, and The Sulphurs —— 243

CHAPTER 36
The Many Camps at Battleship Rock —— 249

CHAPTER 37
Turkeys in the Jemez —— 259

CHAPTER 38
Servants of the Paraclete —— 263

CHAPTER 39
Hippies and Hot Springs in the Jemez —— 269

CHAPTER 40
Sierra de los Valles —— 277

NOTES —— 283

REFERENCES —— 318

INDEX —— 333

PREFACE

This book is a set of short histories of people and places in the Jemez Mountains. Most of these chapters were originally written as articles for the Jemez Springs newspapers the *Jemez Thunder* and *After the Thunder*. I started writing these articles after my wife, Suzanne, and I moved back to the Jemez in 2014. I say "moved back" because I lived here with my family in the 1960s and early 1970s. My father was the district ranger with the US Forest Service, and in 1964 we moved into the ranger's residence across the arroyo from the old mission ruins of San José de los Jémez. I was eight years old when we arrived, and I recall being in a state of awe and wonder in this magical place. In many ways, I still have those feelings.

The Jemez Mountains have always been home, and in some ways I never left. Over four decades of my career as a tree-ring scientist and professor at the University of Arizona, I have returned here often to study forest ecology and land use history in this amazing landscape filled with ancient trees and ancestral Pueblo ruins. After retiring from academia and moving to our place perched below the cliffs of Virgin Mesa, I have had the pleasure of rediscovering places and becoming reacquainted with people I knew growing up. The best part has been that I have learned many new stories that I did not know about before. Still, I feel like I have only scratched the surface of Jemez Mountains history. Many Jemez stories are yet to be discovered, and some are lost to time. That is part of the allure and mystery of the Jemez landscape.

The landscape I refer to here encompasses the entire mountain range but primarily the southern Jemez Plateau. This is the expansive set of mesas and canyons to the south of the Valles Caldera. It includes especially the people and landmarks within the greater Jemez Valley, its headwaters, and its tributaries.

As you will note, the book is an eclectic collection of stories. They are included in five parts grouped more or less by general topic and time, but not strictly so. There is no need to read the chapters sequentially; you can jump around as you please. Because the book is not presented as a linear chronology of the Jemez landscape and may not be read in sequence from beginning to end, there is some repetition of information in chapters, with references to early or later relevant chapters in case you want to follow those topic threads.

There are several recurring voices and content in the chapters. I write in the first person in parts of some chapters and in a memoir style as I recall events, people, and places from growing up here and since then. Several stories relate the experiences of my father, Fred Swetnam, the district ranger. I also draw from my scientific studies of this landscape, especially regarding forests, fires, and human land use history.

Other voices are first-person and newspaper accounts using extended quotes. This is especially so in part 2, with lengthy excerpts from Richard Baxter Townshend's stories of the Jemez in the late 1870s and in 1903. Townshend was a remarkable observer of Jemez people and places, first living here as a young British adventurer and then returning on a nostalgic tourist visit when he was a middle-aged Oxford scholar. Townshend's sensibilities sometimes reflect his Victorian Era prejudices, but his detailed observations and genuine feelings for the people and places of the Jemez provide a rare eye-witness perspective.

Historical comparisons in the book include twenty-three "then and now" photograph sets, including old photos from the late 1800s to the early 1900s. These show both subtle and striking changes that have occurred over the past 100 to 140 years. Old photos and quotes from people also capture a broader theme of the book: the continuity and contrast, the stasis and change, of places and people over time. There is no substitute for the authenticity of voices and photos from times past and present.

The chapters can be read as brief introductions to various cultural and natural histories of the Jemez, and if you want to learn more and dig deeper, be sure to visit the endnotes. The sources for information in the text, photographs, and maps are there, as well as various digressions and side stories. The combination of extensive endnotes, a bibliography, and an index is intended to make this book a reference source for students, historians, and other scholars to learn more about the Jemez.

Regarding my usage of the names "Jemez" and "Hemish:" This place and the people who have lived here for countless generations are most commonly called "Jemez" in modern documents and maps. The early Spanish colonists used multiple different names and spellings, including Xemes, Emexes, Emmes, Jemes, and Jémez. The people's name for themselves is

"Hemish," as described by Joe Sando in his book *Nee Hemish* (2008). Here I use the name "Jemez" for place names, for the people in some historical contexts, or when it is quoted from a document. Otherwise, when I refer to the Jemez Mountains Puebloan people I generally use their name for the tribe, "Hemish."

Finally, in a few chapters I urge acknowledgment and truth-telling about past land uses, land theft, and abuses of people in the valley. I also make suggestions and express hopes for better management of forests by private landowners and government managers. I hope that the chapters about Soda Dam will increase awareness of the uniqueness of this world-class cultural and natural landmark and the need to restore and interpret this place for all.

I hope you enjoy reading this book as much as I have enjoyed writing it. I expect many of these stories will be at least somewhat familiar to longtime residents and Jemez landscape aficionados. Still, I predict that most readers will find new historical revelations and connections in these pages.

ACKNOWLEDGMENTS

I am thankful and grateful to many people who have helped and encouraged me during the research and writing of this book, beginning with my wife, Suzanne. She has patiently listened to countless hours of my telling and retelling of these stories as I learned and wrote about them and shared them with family and friends. I am also grateful that she has forgiven my distraction from and inattention to other matters while I worked on this project.

I thank Robert Borden, editor of the *Jemez Thunder* newspaper, for first encouraging me to submit history articles and for publishing them with various photographs.

I am grateful to archivists, librarians, and others who have been helpful to me in finding research materials and photographs and assisting in obtaining copies and permissions for use. They include Hannah Abelbeck and Catie Carl with the New Mexico History Museum, Palace of the Governors; Diane Tyink of the Maxwell Museum of Anthropology, University of New Mexico; Teddie Moreno and Elizabeth Villa of the New Mexico State University Library Archives and Special Collections; Laura Calderone of the New Mexico State Library; and Chris Morton and Mark Dickerson of the Pitt Rivers Museum, University of Oxford. Thanks go also to Cynthia Gustafson of the New Mexico Department of Transportation for help obtaining documents pertaining to the Soda Dam blasting history; to Pete Taylor, Becky Baisden, and Barbara Zinn for help accessing US Forest Service photographs and historical files; and to Craig Allen for sharing US Forest Service historical documents scanned at the Southwestern Region Office in Albuquerque.

I sincerely thank people who have generously allowed me to use photographs from their personal, family, or institutional collections, or their publications, including Amie Adams, Craig Allen, Matthew Barbour, Barry Kues, Amanda Lewis, Matthew Liebmann, Spencer Lucas, James McCue, Mark Mexal, John Merhege, Gary Morgan, Chris Roos, and David Ryan. Thanks to Tyson Swetnam for creating the Jemez Mountains map with place names.

A special thanks to Miguel and Dee Plana, and Steve Sutherland, who kindly rephotographed many of the "then and now" photo sets that were too

Acknowledgments

difficult for me to do. In most cases, this involved serious hiking and climbing to find the precise spot where the photographers stood 70 to 140 years ago.

I also thank Judith Isaacs for permission to use excerpts from Joe Routledge's stories; Amie Adams for permission to use text from the John A. Adams "San Diego Cañon Grant" document; Robert Cart for stories of his ancestor James Smith; Gilbert Sandoval for stories of Jemez River flooding and Archuleta ancestors; Frank Gonzales for stories about the Forest Service and other topics; John Merhege for Union Oil drilling and Abousleman history memories; Mike Elliott for information on the CCC reconstruction of the Mission San José de los Jémez; and Cathy Stephenson for sharing her memories of the Jemez and clippings of Soda Dam, hippies, and hot springs from her scrapbooks.

I thank the University of New Mexico Press editors, production assistants, and designers for their work on the book, including Michael Millman, James Ayers, Anna Pohlod, Min Marcus, and Isaac Morris. Thanks also to Peg Goldstein for her diligent and perceptive copyediting.

I am especially grateful to people who have read and commented on various chapters and parts of this book, including the earlier newspaper-article versions of some of the chapters. They include Craig Allen, Linda Bedre-Vaughn, Bill Bergmann, Bill deBuys, Fraser Goff, Cathy Goff, Matt Hurteau, John Merhege, Matt Liebmann, Gary Morgan, Bob Parmenter, Janet Phillips, Chris Roos, Cathy Stephenson, Elaine Sutherland, Roger Sweet, Jim Swetnam, and Suzanne Swetnam. I am also grateful to Paul Tosa, Chris Toya, John Galvan, and Ira Sando for teaching me about Towa place names and Hemish history. (Note: Any errors remaining in the text are strictly mine.)

Part I

Stories from Past Centuries

Building stones and timbers of a Jemez field house ruin on Cat Mesa.

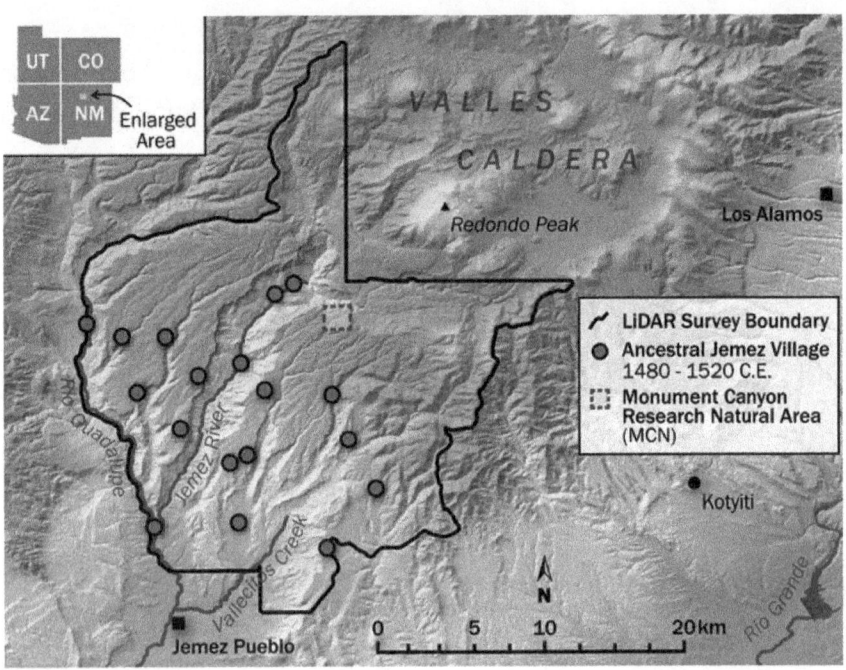

Figure 1.1. Eighteen large ancestral Hemish villages are on the southern Jemez Plateau. This map is from a recent study that estimated the population size of the Hemish at different times using airborne lidar (light detection and ranging) surveys of the village ruin mounds (Liebmann et al. 2016). Today, nearly two thousand Hemish tribal members live in Walatowa.

CHAPTER 1

The Hemish Footprint and Spanish Entrada

The Hemish (Jemez) people have lived in these mountains of north-central New Mexico for a very long time. They lived in San Diego Canyon and on the mesas to the east and west in numerous large villages. You can see one of the smaller village ruins, Gíusewa, on the north end of Jemez Springs, adjacent to the seventeenth-century ruins of the Mission San José de los Jémez. Although the ancient presence of the Hemish here is obvious, I suspect most people are unaware of the extent and scale of their occupation of this landscape for at least 350 years (circa 1300 to 1650).

During this period, the Hemish lived in at least twenty villages with more than three hundred rooms; at least seven villages had more than one thousand rooms.[1] The largest villages were multiple-story "townhouse" structures. By the time of Spanish colonization and the construction of the San José mission church in the 1620s, the Hemish population was probably between six and eight thousand.[2]

In addition to the large villages, the Hemish lived in small one- to four-room structures that archaeologists call field houses. These were mainly inhabited during the summer growing season and were typically near small farm fields. More than twenty-seven hundred of these structures have been mapped, and undoubtedly there are many more that have not been mapped yet.[3]

In a 1946 article titled "The West Jemez Culture Area," Lansing Bloom reviewed the history of the original Spanish explorers and colonists who visited the Hemish ancestral pueblos north of present-day Ponderosa and in San Diego Canyon at Jemez Springs.[4] Bloom was a professor of history

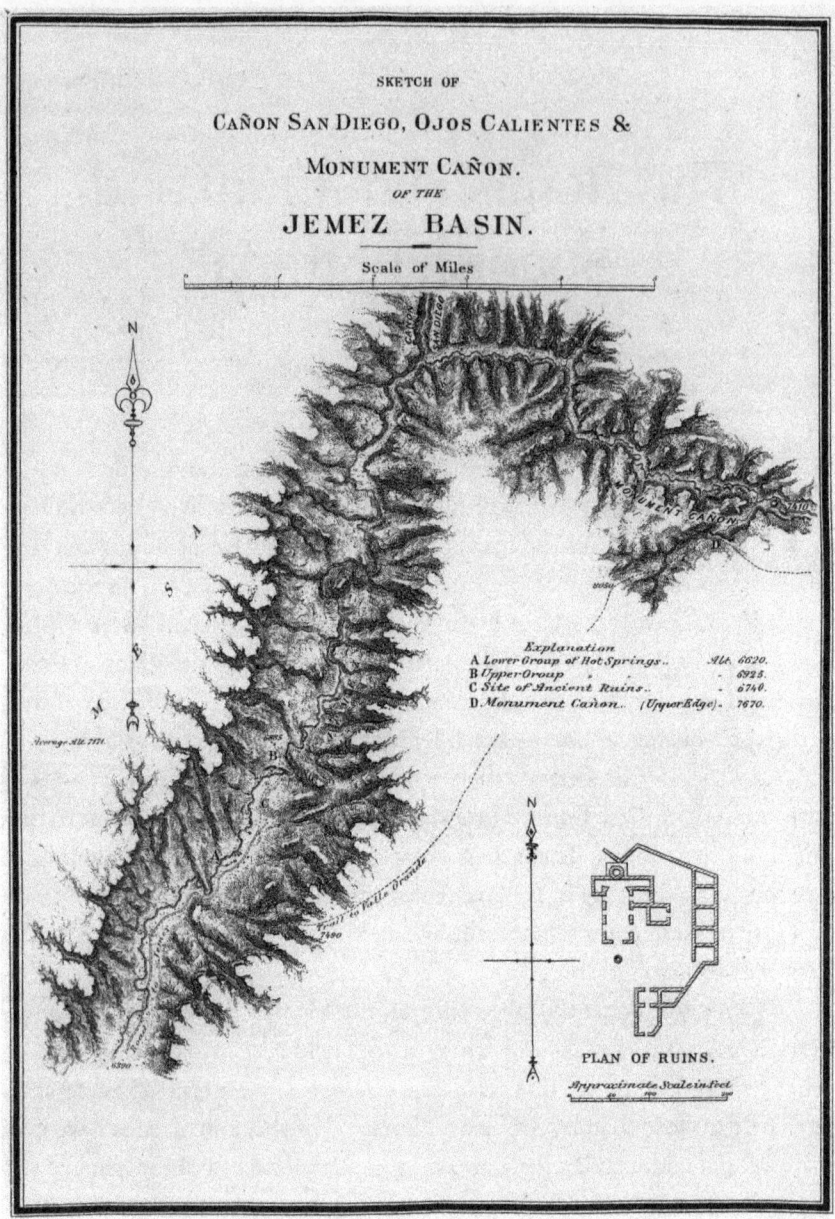

Figure 1.2. This is one of the first maps of San Diego Canyon, from the 1875 geology report of the Wheeler Expedition. The label "sketch" acknowledges that it is a rough approximation of the landscape. A diagram of the San José de los Jémez Mission ruins and a "Trail to Valle Grande" are labeled.

at the University of New Mexico in the 1930s and 1940s. He was involved in the original excavation of Gíusewa and the old mission church in Jemez Springs. In the early twentieth century, he was a leading scholar of Spanish documentary history of New Mexico.

There were several contacts between the Spanish and Hemish people in the mid- to late 1500s, but perhaps most notable was a personal visit by Juan de Oñate to the hot springs and the village of Gíusewa, now in present-day Jemez Springs. Oñate was the leader of the first major group of Spanish colonists in our region. He visited the Jemez in August 1598, apparently traveling down from San Juan Pueblo on the Rio Grande, through the Valle Grande and Vallecitos de los Indios (known as Sierra de Los Piños today), and into the drainages above Vallecitos Viejo (present-day Ponderosa). After visiting the Jemez villages there, he traveled west.

Bloom described his route into the Jemez: "Oñate ... had established his real at San Juan pueblo, and the wording of his report indicates that he had entered the country from the north. He 'descended' through the Valles to the pueblos in the Vallecitos drainage then working to the west over the high mesa land he 'descended' from the potrero to the 'last pueblo' of the province which he associated with the marvelous hot springs. Gíusewa is the pueblo meant beyond any reasonable doubt, and the trail from the Vallecitos down into Hot Springs is still in daily use."

It is most likely that Oñate struck what we now call the Padre Alonzo Trail somewhere on Cat Mesa and then followed it down into Church Canyon to Gíusewa, where the mission church was later built. Since Bloom's paper was written, scholars have concluded that the church at Gíusewa in Jemez Springs was named San José de los Jémez Mission.

The US Forest Service acquired the land the Padre Alonzo Trail traverses in Church Canyon from the Catholic order the Servants of the Paraclete, which purchased it in the late 1940s. I hope the Forest Service reopens that trail for public hiking. It is historic, to say the least. It is a fantastic hike from the old mission to the mesa top or to Monument Canyon, Vallecitos de los Indios, and Jemez Falls.[5] There are several places where heavy use of the trail over centuries is evident, as it is entrenched where it climbs up and over the rimrock of Cat Mesa.

Perhaps the most puzzling part of the history Bloom recounts in his paper is this brief passage: "The pueblo ruin at the Ojo de Chihuahua on the high mesa east of the Vallecito Viejo is not many miles distant through the forest from sites which were occupied by the ancient Cochiteños; and only one and a half miles eastward from that ruin lies a thirty-foot dugout, felled and shaped by the Indians of Santo Domingo and left high on the mountain range like a miniature Noah's ark for which there had been no pressing need."[6]

Isn't that strange? "A thirty-foot dugout." Was this a boat made from a ponderosa pine tree to be floated on the Rio Grande? But why? And why did they leave it? It may be unlikely that a thirty-foot log would still be around after eighty years (since Bloom's paper was written), but who knows? Maybe some remnants of the dugout are still there, if they haven't decayed away or burned up.

Lansing Bloom's description of a dugout on Borrego Mesa and his suggestion that it was an abandoned ark, a watercraft inexplicably high on the mountain and far from a river, puzzled me for months after I read his 1946 article. But recently, it occurred to me the dugout log may have been intended for some other use. Instead of plying the waters as a canoe, it may have been intended for conveying waters as a *canoa*.

In New Mexico, the Spanish term *canoa*, with its obvious English translation "canoe," is also applied to water flumes built from dugout logs to carry acequia waters across arroyos. End-to-end dugout logs are laid down with supporting pillars across gaps in the land to carry lifeblood water to the fields. A picturesque and famous canoa is located near Las Trampas in the Sangre de Cristo Mountains. Maybe Bloom's dugout on Borrego Mesa was intended to be a canoa but for some reason was never dragged from the woods to be used.

Another colloquial variant of the term *canoa* in New Mexico is *canova*.[7] I asked Frank Gonzales, a lifetime resident of Ponderosa, if he knew anything about canoas around Borrego Mesa, and he immediately said, "Oh, yes, there is Canovas Canyon over there." The head of this canyon is in the approximate location where Bloom described the dugout log. So it seems most likely that the dugout log Bloom saw was a canova and the canyon there may have been named after the presence of this abandoned flume or others obtained from there. Perhaps Bloom's canova was to be used near one of the springs around Borrego Mesa or in Vallecitos Viejo (Ponderosa), or maybe it was going to be hauled to Cochiti or Santo Domingo Pueblo. In any case, it goes to show an example of historical mysteries that might be solved and potential connections that are possible when we listen to voices from the past and present.

In this case, I don't exactly know why the water flume possibility occurred to me before I asked Gonzales. It may be because I have seen the Las Trampas canoa. Also, I grew up at the old ranger station in Jemez Springs, across from the old mission ruins where there was a flume carrying acequia water across the Church Canyon arroyo. It was not a log dugout canoa but an early twentieth-century milled lumber and sheet tin–lined flume. As kids we would clamber across the rickety flume, fifteen feet or so above the arroyo bottom, sometimes as floodwaters raged below us!

The Padre Alonzo Trail and the abandoned "ark" (dugout flume) are just two of the many historical and cultural artifacts of the Jemez. These places and objects all have stories to tell. Some of these stories are told in old documents, photographs, and legends of people who have lived here for centuries. Other stories and mysteries will remain unknown and unsolved, to be revealed someday, or not.

In the final chapter of this book, I ruminate on the layering of places and memories associated with them, of people, things, and events, and how the landscape serves to help us remember and make connections.

Figure 2.1. Artist Regina Tatum Cooke's rendition of what the San José de los Jémez Mission looked like when it was completed around 1626 (from Elliott 2011). Cooke was employed by a New Deal program from 1935–1942 to paint Spanish mission churches in New Mexico. (Courtesy New Mexico Museum of Art)

CHAPTER 2

San José de los Jémez Mission

Countless visitors over past centuries have marveled at the red rocks, soaring cliffs, and hot springs as they travel up San Diego Canyon. But another kind of wonder captures their attention as they catch view of the massive stone walls and octagonal tower of the "old mission" ruins. The earliest written accounts of the San José de los Jémez Mission refer to it as a "breathtaking, sumptuous, and distinguished church and friary." Many soldiers, scholars, artists, and photographers who have visited Jemez Springs over the past four hundred years have conveyed the impressive character of the church and ancient ruins in writings, sketches, paintings, and photographs.[1]

The old church is remarkable in its location and appearance, and so are the facts of its creation and history. The existing ruins comprise not one but two churches. These structures were built by Catholic priests and the Hemish people who inhabited the village of Gíusewa, meaning "where the hot waters flow," situated at the mouth of what is now known as Church Canyon.[2]

Two Franciscan priests directed the construction of these churches over short periods. The first was Padre Alonzo de Lugo. Between 1598 and 1601 he served as a missionary among the Hemish and somehow—using a combination of coercion, persuasion, and faith—he convinced the Hemish people to build the first church.

After de Lugo left for missionary duties elsewhere, a Franciscan lay brother continued to minister to the Hemish for some years, and the church may have been partly abandoned. Finally, in 1621 another missionary arrived. Fray Gerónimo de Zárate Salmerón was in his sixties, and he had been a designer and builder of large causeways in Mexico City. Despite his age, he must have been incredibly energetic to have gathered the Hemish at Gíusewa again while at the same time directing them in the construction of the enormous new church. The new church was built alongside the old

Figure 2.2. This photo was taken in about 1910 by Jesse Nusbaum. The house on the east side of the mission was constructed within the old *convento* in about 1866 and appears unoccupied in 1910. Notice the two-story house and barns across the road at the upper left. That was the home of John and Mary Stright Miller and their son, Hugh. They owned part of the property that the mission sat on and they ran a bed-and-breakfast-style hotel in their home for some years. Later, in the 1920s, La Esperanza Hotel (later called the Clay Hotel) was built at this location. After the 1940s, the Via Coeli and Servants of the Paraclete structures were built there. (Courtesy New Mexico History Museum, Palace of the Governors Photo Archives, NMHM/DCA Negative No. 012902)

chapel, which became part of the *convento* rooms on the east side of the church. Zárate Salmerón left the Jemez in 1626.

The Hemish scholar Joe Sando described Zárate Salmerón and the subsequent priest Martin de Arvide as follows:[3]

> According to his own account, Fray Geronimo "sacrificed himself to the Lord among the pagans," toiling chiefly among the Hemish, of whom he baptized 6,566, and in whose language he wrote a doctrina, or theological treatise. According to Benavides, Frey Geronimo was a good priest and linguist, and founded a very beautiful convent and a magnificent Chapel dedicated to Saint Joseph at the principal pueblo of the Hemish (Gíusewa). In spite of the utmost perils and difficulties, then, Fray Geronimo built churches and monasteries and wrote a document [*Relaciones*][4] which, while inaccurate in some details, since he wrote it after he returned to New Spain, is indeed a most important source of information about New Mexico and the Jemez. Today, the ruins of San José Church at Gíusewa are preserved as a state monument.
>
> The magnitude of the construction problems faced by the early mission builders was immense; consider the extremely limited numbers of primitive tools available to them, and the frustrations of trying to supervise helpers whose language was entirely foreign. A sample of the tools issued to the padres when they went out to the mission fields would be 10 axes, 3 adzes, 3 spades, 10 hoes, one medium size saw, one chisel, two augers, one plane, a latch for the church door, two small locks, 12 hinges and about 6000 nails of various sizes.
>
> … In 1628, two years after Fray Geronimo had left, Benavides assigned Fray Martin de Arvide to the Hemish mission field, with orders to revive the two churches and to bring the scattered communities into one area. Fray Martin had served at Picuris Pueblo for many years. In his new assignment, he cultivated land for the Hemish, introducing many new plants. Four years later, after watching the pueblo grow to 300 houses, however, Fray Martin was transferred to Zuni early in 1632; there, most unfortunately, he was killed by the Zunis on February 22nd, only five days after they killed Fray Francisco de Letrado. Soon after his departure, the Church of San José was abandoned and the Church of San Diego became the principal mission, with Walatowa the principal Pueblo.

Figure 2.3A–B. (*opposite*) The old photo was probably taken in the early 1880s. Notice the man standing on the top of the octagonal bell tower. Cornfields, the Stone Hotel, and the bathhouse are visible in the valley. (Courtesy New Mexico History Museum, Palace of the Governors Photo Archives, NMHM/DCA Negative No. 014478)

Figure 2.4. (*above*) The spiral staircase inside the octagonal bell tower has hand-hewn wooden steps notched into a central pole and the walls. The steps were reconstructed by the Civilian Conservation Corps (CCC) in 1936. The CCC also replaced the large pine lintel over the main doorway of the nave. Photo by Stretch Fretwell, 1953. (Courtesy Special Collections and Archives, Cline Library, Northern Arizona University, Catalog No. NAU.PH.98.6.129)

When Fray Martin arrived in 1628, the church may still have been in reconstruction from a fire that occurred during an uprising by the Hemish in 1623.[5] Scholars who have studied the current ruins in detail conclude that the church roof burned but then was replaced. Arvide may have directed the final reconstruction.[6] Imagine the incredible amount of hard work required to cut and haul huge roof timbers to span the fifty-foot-wide nave and walls, hew them into shape, and then hoist them up onto the thirty-four-foot-high walls over the nave—and to have done that twice within a decade!

These early years of the seventeenth century were very turbulent in the Jemez Mountains, with frequent raids by Navajo. There was also anger and dissatisfaction among the Hemish with the heavy taxes and harsh punishments of the secular Spanish alcaldes and governors. This led to intermittent uprisings by the Hemish and other tribes, such as in 1623, ultimately leading to the coordinated Pueblo Revolt across Northern New Mexico in 1680, in which the Hemish played an important role.[7]

The historical record suggests that San José de los Jémez may have been abandoned by the priests or used only intermittently as a church after Fray Martin left in 1632. The Hemish continued to use the site in the mid-1600s, and they constructed an aboveground kiva near the convento complex. The village and church were probably no longer occupied after the Pueblo Revolt of 1680. However, someone built a house in the old convento in about 1866.[8]

My family moved to Jemez Springs in 1964, when I was eight years old. Our house was the old US Forest Service ranger's residence, located about 150 yards south of the mission ruins, across the arroyo of Church Canyon. I

Figure 2.5A–B. (*opposite*) These photos of the excavated and artist-reconstructed decorated wall of the nave show the fleur-de-lis (lily flower). The symbol was used by the Catholic Church to represent the Holy Trinity or some combination of purity, the Virgin Mary, and Saint Joseph. The latter might be particularly relevant in this case, given the name of this church. (From Elliott, National Historic Landmark nomination form)

recall being in a state of absolute wonder and awe of this amazing place we had moved to!

At that time, the state monument visitor center had not been built yet, and there was no caretaker there. For me, my three siblings, and village kids, the old mission was our backyard playground, fort, and climbing place. We roamed freely without restriction; we were "free-range kids." We climbed up and down over all the walls, no doubt knocking wall stones off here and there, contributing to the ruin mounds below.

Our favorite destination at the old mission was the top of the octagonal bell tower. To get up there we would scale the north wall of the tower on the hillside and then sidle around on a very narrow ledge on the east side of the tower to the front. From there we could climb through the vertical opening and step onto the rickety spiral staircase, which wound around a pole to the top. I recall the wooden steps as appearing hand-hewn and being mostly rotten. We avoided stepping on the rotten ones, and somehow they held our childish weight to the top. We often took picnic lunches with us and enjoyed the view with our peanut butter sandwiches.[9]

It was a sad day for us kids when the visitor center was opened and dedicated by New Mexico governor Jack Campbell in August 1965, because the new caretaker and manager moved in and we were banned, of course, from climbing on the walls or up the bell tower. Fortunately, the state has continued to stabilize and protect the ruins from kids and other natural forces of erosion over the past sixty years.

There are many wonders to learn about and mysteries yet to discover in the old mission. For example, when the rubble, which was up to ten feet deep, was removed from the nave in the 1930s, archaeologists discovered frescoes with fleur-de-lis motifs on both east and west walls. This was a rare find among the early Franciscan missions in New Mexico.[10]

Also, when you visit next time, notice the shape of the window on the west wall, opposite the altar. This "splayed window" was designed like those in castle walls, with acute angles on the vertical sides, allowing defending archers or riflemen to shoot at attackers down the length of the walls. As Adolf Bandelier said when he described the ruins and bell tower in 1891, the church was "manifestly erected for safety and defense."[11]

CHAPTER 3

Pueblo Revolts and Battles of Reconquest

The great Pueblo Revolt of 1680 was the most successful defeat and expulsion of European colonists by Native Americans from a region of North America. The Jemez and Zia people played key roles in the revolt, especially the events that followed during the multiple attempted reconquests by the Spanish, leading up to three pitched battles in the Jemez Valley in 1689, 1694, and 1696. Stories of those battles come to mind almost every time I drive up the valley past Zia and Jemez Pueblos, and when I see the striking red and buff cliffs of Guadalupe Mesa (known in Towa as Munstiashinkiokwa).[1]

The history of the Pueblo Revolt and the reconquest are detailed in definitive histories by Joe S. Sando (*Nee Hemish*) and Matthew Liebmann (*Revolt*).[2] I highly recommend those books for anyone curious about the details of this remarkable history. Given these excellent sources and others, I recount here only a brief synopsis with sketches of events, places, and people during the reconquest period (1692–1696) that stand out as particularly amazing to me.

These descriptions are based on documentary accounts and other facts. But in some ways they are so fabulous that if they were written in a novel, you might find them implausible or, as the saying goes, "stranger than fiction." Some events are pure legend but are vividly remembered and told over the centuries, with emblematic images on mesa walls and commemorations carved on boulders.

One of the most remarkable historical figures during the reconquest period was Bartolomé de Ojeda.[3] He was a mestizo (sometimes also known as a coyote in New Mexico), with Spanish and Zia parentage.[4] In addition to his mother tongue of Keres, he was fluent in Spanish and possibly also

Figure 3.1. The view north, via drone, up Guadalupe and San Diego Canyons, with Guadalupe Mesa in the center. The village of Patokwa was on the low snowy bench in the center left foreground. (Courtesy James McCue)

Towa and Tewa languages. He had been baptized and educated by Franciscan priests and could read and write in Spanish. His role during the 1680 Pueblo Revolt is not clear, but he was probably a war captain on the Pueblo side, as he was in 1689.

The written accounts of his life begin during the attempted reconquest in 1689 by Domingo Jironza Petrís de Cruzate. In a fierce battle at Zia Pueblo, Ojeda fought bravely against the Spanish, even though wounded by a musket ball and an arrow. More than six hundred Puname people (Zia, Santa Ana, and San Felipe) were killed in this battle, and seventy were taken captive.[5] As the story goes, when the Spanish approached the gravely wounded Ojeda, he called out to a Franciscan friar to hear his confession and give him the last rites. Despite his wounds, he survived and was taken captive to El Paso as the Spanish retreated from the failed reconquest.

Over the next few years, the recovered Ojeda became a primary informant to the Spanish on what had transpired in New Mexico during and following the 1680 Revolt, including gruesome descriptions of the killing of seven priests. His testimony served to enrage the Spanish and harden their resolve to retake New Mexico. He also talked about infighting between the Pueblos and with "Apaches" (used in old Spanish documents to mean both Apache and Navajo). This news revealed that some tribes no longer followed Popé, the leader of the 1680 Revolt, so they might be open to alliance with the Spanish. By 1692 Ojeda was considered a valuable ally, and he accompanied Diego de Vargas and his army to New Mexico. Over the subsequent years of the reconquest, he served as a key guide, interpreter, and intermediary with the Pueblos.[6]

After their devastating defeat in 1689, the surviving Zia people moved to Cerro Colorado, an isolated and highly defensive mesa due west of what today is the Hemish village of Walatowa. It was here that Ojeda was sent by Vargas to convince the Zia to surrender and join the Spanish as allies in the reconquest. It must have been a stunning scene as Ojeda limped up the trail to the refugee village on the mesa, carrying a cross in one hand. When he came closer, they recognized their former war captain!

Ojeda was a bold and persuasive man who had credibility and some trust among the Zia. The Zia ultimately agreed to join the Spanish, and as a demonstration of their conversion, many were baptized. Subsequently, Zia served as auxiliary fighters in multiple Spanish-Pueblo skirmishes and battles during the reconquest period, including against the Hemish.

As New Mexico historians Richard and Shirley Flint wrote:

> With the assistance of Ojeda and other cooperative Pueblo intermediaries, Vargas was able to perform similar rituals of submission and possession at the other Rio Grande pueblos. By January 1693, Vargas was once again in El Paso, preparing to take up residence in Santa Fe later that year. That reoccupation would prove much more difficult and violent than had the tour of 1692. But unflinchingly committed to the benefits of life under Spanish rule, Bartolomé de Ojeda would be Vargas's dependable ally throughout the conflicts that ended organized Pueblo resistance to that rule.

After the people of Zia had formally submitted to Spanish sovereignty, Ojeda took up residence there. Having been educated by Franciscans as a child, Ojeda was literate in Spanish. Using that skill, he communicated with Governor Vargas by letter whenever urgent matters presented themselves. In January 1694, for example, Ojeda warned Vargas that the people of Jemez were planning either to attack the Spanish horse herds or to abandon their pueblos and flee from the area of Spanish control, to live with Apache Indians beyond the frontier. He repeatedly asked the governor for military assistance to defend Zia against hostile neighboring Pueblos. As at the mesa of La Cieneguilla, where the Cochiti people had taken refuge, Ojeda provided reconnaissance information to Vargas. In that case, he even led the key assault on the mesa from the rear resulting in the capture of 342 Pueblo people.[7]

Ojeda's warning letters and requests for assistance against the Hemish finally convinced Vargas to march a force of about 120 Spanish soldiers and militia, and another 100 Puname warrior allies (Zia, Santa Ana, and San Felipe) to the foot of Guadalupe Mesa in July 1694. After the 1680 Revolt, the Hemish had left Walatowa and reoccupied an old village called Patokwa, located on the low red sandstone bench just above the junction of the rios Guadalupe and Jemez. They also built a new village, Astialakwa, on top of the high Guadalupe Mesa directly above Patokwa.

Interestingly, refugees from other Pueblo tribes and Navajo allies also joined the Hemish at the Astialakwa stronghold. The high village was accessible mainly by a narrow, steep trail on the southern end. The village rooms were stocked with sufficient corn and provisions to withstand a long siege by the Spaniards.

But the siege was not long; the Battle of Astialakwa was fought on a single day, July 24, 1694. The Spaniards and their allies used a pincher tactic, with separate attacks on the south and north sides of the mesa. In addition to the established trail on the south side as a route of attack, Ojeda had provided key information about a second attack route ascending the north side from the saddle between Guadalupe and Virgin Mesas.

One story has it that among the Pueblo "allies" who had joined the Hemish on the mesa at Astialakwa was a group of infiltrator spies, who at a

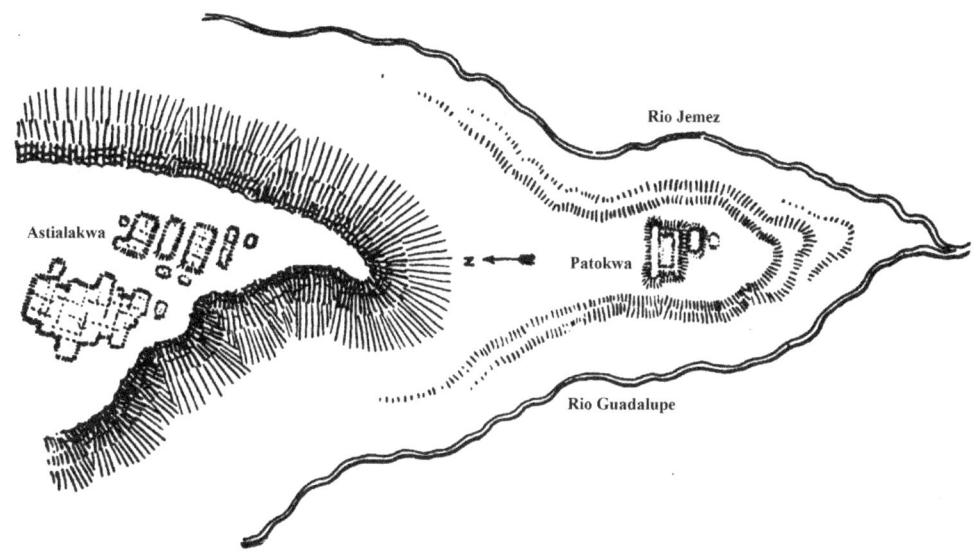

Figure 3.2. Map of Astialakwa on Guadalupe Mesa and Patokwa at the junction of the Jemez and Guadalupe Rivers. (Adapted from Holmes, "Notes on the Antiquities of the Jemez Valley, New Mexico")

crucial moment turned on the Hemish and helped the attackers overwhelm the defenses.[8] What is quite striking about this story, and the overall history of the reconquest, is that the fighting was not simply Spanish against Pueblo. The mixed-blood Ojeda exemplified this, as he was one of the leaders directing Pueblo auxiliaries (Puname) in the assault on the Hemish and their allies at Astialakwa that day in 1694.

The attackers reached the mesa top and proceeded to reduce the village, burning the houses and killing dozens of warriors and villagers. At a climactic moment, some of the warriors—according to both historical accounts and legend—leaped from the cliffs, choosing death on the rocks below rather than capture and probable execution or maiming by their enemies. The legend then becomes the most fantastic part of the story: As the warriors fell through the air to their certain deaths, San Diego (or another saint, perhaps Santiago) appeared, and the falling warriors were gently levitated to the ground. To this day, an image of San Diego commemorating the event can be viewed on an east-facing cliff of Guadalupe Mesa.[9]

According to various accounts, about eighty-four Hemish and their allies were killed in the fighting, and seven warriors jumped (to their deaths).

Figure 3.3. Navajo artist Charles Keetsie Shirley's drawing, from a 1942 *Desert* magazine article, depicting Spanish soldiers on Guadalupe Mesa, with the burning of the Astialakwa village in the background.

At least eighty-two survived or retreated down the steep scree slopes of the mesa. They went into the mountains, probably to some of the Hemish ancestral village sites, and later (after the battle of 1696) others went as far as the Chama River valley, where they built highly defensive refugee structures. Another group went all the way to the Hopi villages, where they stayed as guests, some not returning to Walatowa until 1716.

The remaining 361 Patokwa/Astialakwa villagers were taken hostage by the Spaniards. They were mostly women, children, and the elderly. They were made to haul a huge quantity of dried corn they had stored in preparation for a long siege down from the high mesa. This back-breaking work, ascending and descending the steep trail with heavy loads, took sixteen days.[10] Vargas had ox-drawn carts brought from Santa Fe, and 420 *fanegas* (more than twenty tons!)[11] of corn were loaded and hauled to Spanish-allied pueblos and back to the capital. The hostages were taken there too, and word was sent to the dispersed Hemish that the return of hostages depended upon surrender. Finally, after months of captivity, the people were released in September, and both villagers and some of the dispersed Hemish warriors resettled at Patokwa.

At this time, amazingly, another Franciscan priest took up residence at Patokwa along with a small contingent of soldiers. They directed the

Figure 3.4. A 1942 *Desert* magazine article quotes an old Navajo medicine man: "Four 'old-men's-lives-ago' the great grandmothers of a Navajo family dwelt there. Twenty 'arrow-flights' north of the pueblo is a deep saddle in the mesa. This Navajo clan took its name from this place, Ma'iidesgizh, Coyote Pass." Other accounts confirm that Navajo allies joined the Hemish at the Astialakwa stronghold. Some of the Hemish refugees from the 1694 and 1696 battles lived with Navajo clans for years. This Shirley drawing from the 1942 article is captioned, "Some of the women and children got out of the trap. Hurriedly they ripped off their woven belts and buckskin leggings. One end was tied to a tree and the other dropped over the ledge. One by one the women slid down the crude rope. On the rim Spaniards shot at everything that moved."

construction of a small church, which became known as San Diego del Montes y Nuestra Señora de Remedios. The peace didn't last long. On June 4, 1696, the Hemish rebelled again, killing the priest at the new church as well as four Spanish soldiers stationed there.

This new rebellion led to the final battle of 1696. Diego de Vargas's forces arrived on June 29 to put down the third revolt of the Hemish within two decades. This time the fighting occurred along the Jemez Valley floor, with skirmishing and all-out warfare at various places, from near Patokwa to the Red Rocks area near present-day Walatowa. The final defeat of the Hemish occurred there. But they did not surrender; instead, the survivors dispersed to the mountains and elsewhere.

The next two decades were a period of diaspora for the Hemish, and for a short time virtually all the traditional Towa speakers left the southern Jemez Plateau, traveling to locations largely unknown. In 1696 Vargas reported the Hemish were "absent." We know that many were in refugee sites to the west and north of the Jemez and as far away as the Hopi villages.

By 1703 some Hemish began to return to both Patokwa and Walatowa, and by 1706 three hundred Hemish were said to reside at San Diego de los Jemez (Walatowa). In 1716, 113 Hemish were forcibly removed from the Hopi villages in Arizona and returned to Walatowa. Finally, sometime in the mid-eighteenth century, the remaining Hemish at Patokwa left that village and coalesced with the people at Walatowa.[12]

Undoubtedly, the sounds of booming muskets and cannons, and the shouts and screams of the warriors, soldiers, and villagers, echoed for generations in the memories of survivors. What remains today are the places, the stories, and the images in stone that represent them. In addition to the profiles of San Diego and the Virgin of Guadalupe on cliff faces, there are petroglyphs—carvings in the rock—that evoke those times and people. These include a large carved image of a warrior with a bow and arrow near the trail and summit leading up to Guadalupe Mesa. Another is an enigmatic carving on a boulder in San Diego Canyon below the mesa, with two coyotes, multiple lines, and circles.[13]

Figure 3.5. This image of San Diego (or Santiago), with long hair, beard, and flowing robes, appears on an east-facing cliff near the north end of Guadalupe Mesa. The inset photo at lower left shows a view of the location of the image from NM 4, between mileposts 11 and 12.

CHAPTER 4

Treasures of the Spanish Queen Mine

Visitors to the Jemez Valley might wonder why a picnic area located about three miles below Jemez Springs is called the Spanish Queen. Longtime residents of the valley know it is named for the nearby Spanish Queen Mine. Legends of fabulous gold, silver, and copper taken from this spot stretch back to the days of the conquistadores. Although tales of great riches obtained from the red rocks of the mine are probably mythical, low-grade copper deposits have been dug out and smelted there for centuries.

The Spanish missionary Gerónimo de Zárate Salmerón, who supervised the building of the Mission de San José at Gíusewa (Jemez Historic Site), was also a prospector. He was probably the first to dig ore at the Spanish Queen Mine. In late 1629 he said, "In all the ranges of the Hemex [Jemez] there is nothing but deposits, where I discovered many and filed on them for His Majesty. From the which I took out eighteen arrobas [456 pounds] of ore."[1]

Promoting the mineral potential of the Jemez Mountains in 1634, Fray Alonso De Benavides, said: "Their [Jemez] pueblos were founded among some terribly rugged, uninhabitable mountains, although extremely rich and prosperous in deposits of silver, and in particular in the finest copper ever seen, since gold is extracted from it."[2]

Figure 4.1A–B. (*opposite*) The stone wall ruins in the 1905 photo were just north of the Spanish Queen Mine and were probably structures related to early mining at this site. The second photo was taken in 2022. (1905 photo courtesy New Mexico State University Library Archive, Special Collections, Image No: 01040329)

The first mention of the mine in English-language documents is by Lieutenant James Simpson of the US Army. On August 20, 1849, Simpson and a few members of his reconnaissance from Santa Fe to the Navajo Country traveled up San Diego Canyon to visit Los Ojos Calientes (Jemez Springs). Along the way, he noted, "Nine miles up the cañon we found an old copper-smelting furnace, which looked as if it had been abandoned for some considerable period. It is quite small, is built of stone, and has arched ovens traversing each other at right angles, each oven being furnished with a stone grating. We picked up some fragments of copper ore [probably green malachite] which lay scattered around."[3]

Later in his account, he said, "At San Ysidro I called to see Señor Francisco Sandoval, the proprietor of the copper furnace we saw two days since up the Cañon de San Diego. He informs me that the mine near this furnace was worked until about three years since [about 1846]; that one man could get from it ten arrobas [253 pounds] of rich ore per day, and that gold was found in association with it. He further stated that he had now cached near the furnace twenty-three arrobas [582 pounds] of pure copper."[4]

It is remarkable to think about arched stone furnaces that were used to smelt copper out of the red rocks in this spot. We are left wondering what happened to those structures and where the stones are now. Were they used in later years in the walls of old buildings in Jemez Springs? Stone wall ruins and remains of other mining structures were still visible around 1900, and some are still near the old mine entrances. If you know where to look, the waste tailings of three of the four horizontal tunnels (adits) are still visible through the cottonwood trees while driving on Highway 4 past the mine.

The latest workings of the mine were from 1924 to about 1937. According to a US Bureau of Mines report, the Burnett Mining Company acquired the property in the 1920s and in subsequent years shipped processed ore to smelters elsewhere. Considerable processing work was also done on the site with equipment located there, including a mill, a leaching plant, and a small smelter. A New Mexico government mining report documents these production amounts in two years of operation: 1928, twenty-nine ounces of silver and fifty-five hundred pounds of copper; 1929, one ounce of gold, ninety ounces of silver, and ten thousand pounds of copper.[5]

The main adit of the mine was purposely collapsed many years ago by dynamite blasting, and there is no sign today of the small opening that

Chapter 4

wild Jemez Springs kids used to crawl through in the 1960s. We ventured down the dark tunnels with flashlights. I recall one very scary-looking vertical shaft within the deepest adit, extending downward into the blackness where our flashlights couldn't reach. Old, decayed timbers and boards held up the sides, with a rickety-looking ladder going down. We were foolish kids but not crazy enough to try climbing down into that black hole!

Gold, silver, and copper are not the only treasures of the Spanish Queen Mine. Another kind of treasure was discovered there in the red rocks, and these discoveries continue today. In 1930 a Harvard University professor of paleontology, Alfred Sherwood Romer, was studying ancient fossils in red beds of the Chama River valley. He decided to look for more fossils elsewhere in New Mexico in similar red sandstones laid down during the Permian period. He came to the Spanish Queen Mine, which was operating at the time, and he immediately hit paleontological pay dirt.

Romer found numerous fossils of animals near the mine preserved in the mudstones and sandstones of the early Permian Abo formation.[6] Other scientists later found many plant fossils within the roof and walls of the mine tunnels. These fossils are about 290 million years old. The red rocks area of that time was a vast floodplain with meandering rivers, large seed ferns, broadleaf plants, and tall conifers. The largest land animals on Earth at the time roamed this landscape as both predator and prey. One of the animal fossils Romer found was a *Sphenacodon ferocior*, a lizard-like predator up to nine feet in length.

In subsequent decades, and especially in the past twenty years, paleontologists have continued to find more animal and plant fossils in the red rocks from the Spanish Queen Mine area south to the junction of the Guadalupe and Jemez Rivers. Spencer Lucas of the New Mexico Museum of Natural History and his colleagues with the Carnegie Museum of Natural History in Pennsylvania and the Smithsonian Institution in Washington, DC, have documented eighteen locations with these fossils.[7]

One of the more important fossil finds is of the *Dimetrodon occidentalis* species, another large lizard-like animal with a tall dorsal "sail." The Spanish Queen fossil site established with incontrovertible evidence that the *Dimetrodon* species existed in northern New Mexico, not just in Oklahoma and Texas, where it was first found. The co-occurrence of *Sphenacodon* and *Dimetrodon* is also a rare find.

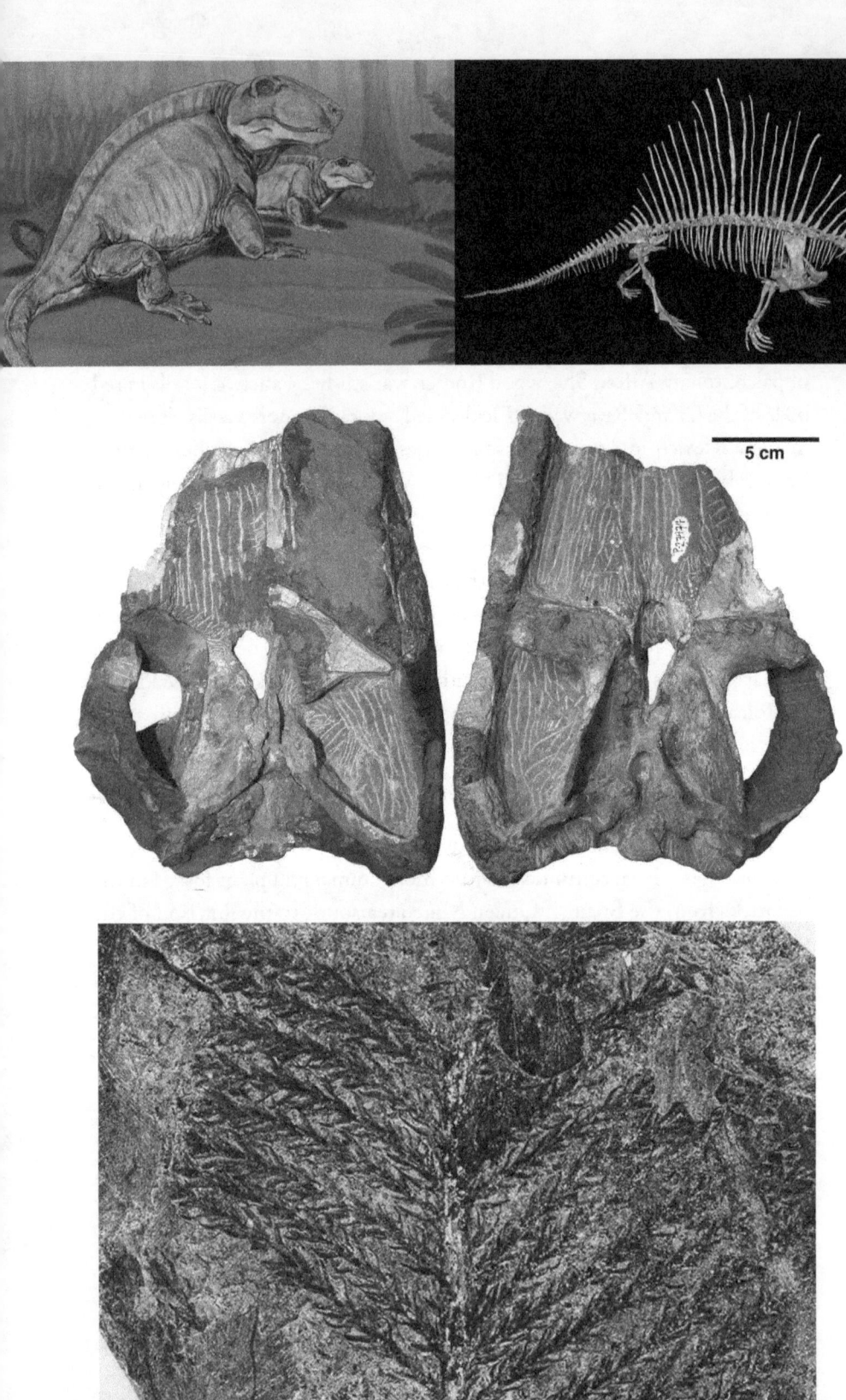

Figure 4.2A. (*opposite top left*) Artist drawing of two *Sphenacodons*. (From Wikimedia Commons, https://commons.wikimedia.org/w/index.php?curid=2461259)

Figure 4.2B. (*opposite top right*) *Dimetrodon* skeleton. (From H. Zell, GNU Free Documentation License, https://commons.wikimedia.org/w/index.php?curid=12214271)

Figure 4.2C. (*opposite middle*) Partial *Sphenacodon* skull fossil found in Jemez Valley red rocks near the Spanish Queen Mine. (From Lucas et al., "Lithostratigraphy")

Figure 4.2D. (*opposite below*) Fossil conifer leaves and a branch of *Walchia piniformis*, collected from the Spanish Queen Mine site. (From Lucas et al., "Lithostratigraphy")

Figure 4.3. (*above*) *Autunia naumanii* fossil from the Spanish Queen Mine. It was a large treelike plant with a cupule of modified leaves containing seeds. (From Lucas et al., "Lithostratigraphy")

These animals were originally thought to be early reptiles, but in recent decades paleontologists have learned they were in a separate line that led to mammals. The dinosaurs came from another evolutionary line, and they appeared in great numbers and sizes after the massive Permian extinction about 252 million years ago.

With this history in mind, hopefully your next drive through the red rocks of our Jemez Valley will be enriched in multiple ways by stories of gold, silver, and copper, with visions of conquistadores, miners, and giant lizards.

CHAPTER 5

Hillers and the Stevensons in the Jemez

The photograph is dated 1879, and it was taken by John "Jack" Hillers. It may be the oldest well-dated photo taken within the Jemez Valley. Hillers, a photographer, accompanied Major John Wesley Powell on his second expedition (1872) floating down the Colorado River in small wooden boats.[1]

In 1879 Hillers was a photographer for a Bureau of American Ethnology (BAE) expedition to New Mexico led by James Stevenson and his wife, Matilda "Tilly" Coxe Stevenson.[2] The BAE was created by an act of Congress to collect historical and cultural materials relating to the Indians of North America and transfer them to the Smithsonian Institution. Objectives of the Stevenson-Hillers expedition (1879–1881) were to photograph the pueblos of the Southwest, to purchase cultural objects, and to conduct ethnographic interviews of the Pueblo people. Hillers's photos are an amazing record of ancient ruins, pueblos, and people at that time.[3]

The sharp details of Hillers's old mission photographs are remarkable. They were taken with a large bellows camera on a tripod, and the negative was a glass plate. In expert hands these cameras produced photos with amazing depth of field; objects in the far distance are nearly as sharp as those in the near foreground. For example, look closely and you may see three people standing on the ruined wall of the church in the distance. And notice the details around the small pool of water in the arroyo at the bottom of the image.

Also perhaps of interest: The trickle of water in the foreground is the ephemeral creek of Church Canyon. The bridge adjacent to the current Jemez Historic Site crosses that drainage, which is now several times deeper than it was in 1879. Floods down Church Canyon during the past century-plus, especially in the 1880s and 1890s, cut the deep arroyo there.

Figure 5.1 Mission de San José in Jemez Springs. Photo by John Hillers, 1879. (Courtesy New Mexico History Museum, Palace of the Governors Photo Archives, NMHM/DCA Negative No. 109999)

Figure 5.2. Portion of Hillers's 1879 photograph showing a couple in the foreground, probably James and Matilda Stevenson.

The Stevensons were a remarkable couple. They were probably the first husband-wife scientist team working for the US government. James had been a staff officer in the Union Army during the Civil War, and in the 1870s and 1880s he was a chief executive for Major Powell, who was the first director of the BAE and then the second director of the US Geological Survey. James was self-taught as a geologist, and he assisted in several government explorations and mapping surveys of the western United States.

Tilly had studied mineralogy, but after marrying James and working as an assistant at the BAE, she became interested in ethnology. A subfield of anthropology, ethnology is the study of the characteristics and cultures of various peoples, and the differences and relationships between them. Tilly focused most of her career studying and writing about Pueblo cultures in New Mexico, especially those of Zuni and Zia Pueblos.[4]

James and Tilly visited the Jemez again during an 1887 expedition with Major Powell and other scientist luminaries from Washington, DC. They camped for two months near what is today Hummingbird Music Camp.[5]

James died in 1888 from a case of Rocky Mountain spotted fever, a bacterial infection transmitted by a tick bite. Tilly continued her work with the BAE and later founded the Women's Anthropological Society in Washington, DC. Women were not allowed to be members of other science societies of the day. She overcame many hurdles as a woman scientist working

Figure 5.3. (*opposite above*) Hillers's photo of the old mission church in 1879 or 1880, taken from the hill above the Church Canyon arroyo to the south. The photo shows a well-established home within what was once the *convento*. Notice the *horno* (Pueblo-style outdoor oven) and part of the ruins used as a livestock pen. A photo from the 1920s or 1930s shows the same home with a pitched roof. (Courtesy New Mexico History Museum, Palace of the Governors Photo Archives, NMHM/DCA Negative No. 016523)

Figure 5.4. (*opposite below*) This photo from a 2019 *New Mexico Magazine* article is captioned, "Curator Betty Toulouse (*right*) holds sacred Zia pot that inspired Reba Mera's (*left*) state flag design, c. 1940–50." (Courtesy New Mexico History Museum, Palace of the Governors Photo Archives, NMHM/DCA Negative No. 050131)

in a male-dominated world of science and government. She was not treated well, but she persevered.[6]

In later years, an accusation darkened her reputation, perhaps unfairly. In about 1890, after Tilly had visited Zia Pueblo, a sacred pot from the tribe's Fire Society disappeared. To this day, no one knows how the pot was removed from its kiva, but some suggested that Tilly had taken it. Her possible involvement with the missing pot was never proven, and the pot wasn't in the Smithsonian collections.

The Zia sun symbol on the pot may have been a source of the image appropriated by the state of New Mexico and many others since. A century later, the pot showed up in a private collection in Santa Fe and was returned to Zia Pueblo in 2000.[7]

CHAPTER 6

Myths and Legends of Montezuma in New Mexico

You probably know that Montezuma was the great chief of the Aztec. The Spanish conquistador Hernán Cortés tricked and murdered him in Tenochtitlan (Mexico City) in 1520, when the Spanish brutally conquered ancient Mexico. But do you know the legend that Montezuma was born in New Mexico and was magically transported to Tenochtitlan, where he assumed his role as emperor of the Aztec?

The Montezuma–New Mexico mythology was at a peak during the mid-1800s to early 1900s. This probably explains the first verse of "O Fair New Mexico," selected as the official New Mexico state song in 1917:

> Under a sky of azure, where balmy breezes blow,
> Kissed by the golden sunshine, is Nuevo México.
> Home of the Montezuma, with fiery hearts a glow,
> State of the deeds historic, is Nuevo México.

There are many different Montezuma legends in the Southwest, with different storylines told and retold among the Pueblos, Spanish, and Anglos. Some of these stories are still told today in various forms, such as during the Pueblo Matachines dances, with characters representing El Monarca (Montezuma) and La Malinche (a Native or mestizo woman, Montezuma's wife or daughter, or the wife of Cortés, depending upon the story).

A remarkable historical analysis of the Montezuma myth in New Mexico was written by the great Swiss scholar and ethnographer Adolf Bandelier in 1892.[1] Bandelier spent many years studying original Spanish documents in the Southwest, Mexico City, and Spain. He concluded that the myth was imported to New Mexico, with a major elaboration and dissemination, just before the American invasion of Mexico in 1846.

Bandelier outlines multiple possible early origins of the New Mexico Montezuma legend, including old migration tales of people from somewhere in North America moving southward to Mexico. The migration of people from the north to Tenochtitlan and later the return of their descendants during the Spanish Entrada into New Mexico—which included many mixed-race Indigenous-Spanish people—is at the heart of the Aztlán mythology. That mythology had a resurgence of interest in the 1960s and 1970s with the development of an Indo-Hispanic Chicano movement.[2]

An interesting part of this mythology and history is Bandelier's theory that a mid-nineteenth-century version of the New Mexico Montezuma legend was a Mexican government propaganda ploy. Bandelier describes a "History of Montezuma" manuscript, "never printed," which he claimed existed as multiple copies in New Mexico. He recounts how the document placed Montezuma's origins in New Mexico, relating his childhood history to legends of Pueblo "hero gods" who rose to greatness and then left but promised to return.

He also says the story connects La Malinche to New Mexico:

> The document cited makes of the Malinche a daughter of Montezuma, and, after bringing in Cortes and his conquest and victory over Montezuma, concludes by marrying the Malinche to Cortes, and by representing New Mexico as a part of the dower [sic] which the Indian maiden brought to her Spanish husband. Such a document, manufactured at a time when an American invasion of New Mexico was apprehended, written at the City of Mexico and circulated in every New Mexican pueblo that could be reached, is plainly what may be called a "campaign document," conceived in view of strengthening the claims of Mexico upon New Mexico in the eyes of the Pueblo Indians and refuting anything to the contrary that might be anticipated from the side of the United States.[3]

Although Bandelier argued the document was probably written in Mexico City, he acknowledged the possibility that it was written by someone in New Mexico. In support of this alternative, he refers to an old printed

Chapter 6

book at Jemez Pueblo that his friend Archbishop Jean-Baptiste Lamy saw there and partially transcribed. He shared the transcription with Bandelier, who concluded it was a copy of the "Letters (Cartas) of Cortés edited by Lorenzana [1776] and illustrated with pictures of Mexican costumes, etc."

Bandelier suggests that this book could have provided the "material" for someone in New Mexico to write what he characterized as a "campaign document." He says, "The existence of which [this book] was known to all the Pueblos and about the contents of which they had been partially informed, it would have been easy to gather material for the 'History of Montezuma' of 1846." However, the lack of subsequent confirmation of the existence of handwritten copies of the "History of Montezuma" led a later scholar to describe Bandelier's theory as "shrouded in doubt."[4]

Bandelier's 1892 article was probably a response to a popular and widespread retelling of a version of the legend in 1885 by the former territorial governor of New Mexico William G. Ritch, an Anglo, in a lavishly illustrated book, *Aztlán: The History, Resources, and Attractions of New Mexico*.[5] Nearly one hundred thousand copies of this book, including earlier editions under the title *Illustrated New Mexico*, were distributed nationally. It was intended as a promotional vehicle to bring settlers and capitalist investors to New Mexico.

The image shown here of Montezuma flying on the back of an eagle from New Mexico to Mexico City is from that book. Notice that Santa Fe and Mexico City are labeled on the continent below the clouds. It is unknown how much of Ritch's version of the legend is from the "History of Montezuma" document and what parts he completely made up. Ritch's intentions in retelling his version of the legend are not entirely clear, but a scholar recently interpreted it as an American Dream–like appeal; a glorification of individualism portraying the rise of a poor person, born in New Mexico, to greatness and riches.[6]

Here is Benjamin Read's summary of Ritch's telling of the legend, with the final iconic Mexican imagery:

> A written record which is to be found in some of the Pueblos is that Pecos pueblo was the birth-place of Montezuma; that after he had grown to man's state he showed himself possessed of supernatural powers; that he at a certain time assembled a large number of his people and started from New Mexico on a journey south, Montezuma

41

Figure 6.1. Montezuma riding an eagle from New Mexico to Mexico City. (From Ritch, *History, Resources, and Attractions of New Mexico*)

riding on the back of an eagle; and thus riding in advance, as to his people as was the star to the wise men of the East. The sign of arriving at the site of the great city and capital of the Aztec nation was to be the alighting of the eagle upon a cactus bush and devouring a serpent. This event took place when the eagle arrived at the site of the present city of Mexico, then first made a city and capital.[7]

Aztlán may be more of a concept than an actual place, but it is evident that there were indeed ancient cultural ties and trade between New Mexico's Pueblo people and Mesoamericans. Archaeological findings at Chaco Canyon show that trade with people far to the south existed more than one thousand years ago. Plants, animals, and artifacts that must have come from deep in

Figure 6.2. Matachines dancers from Jemez Pueblo at Our Lady of Assumption annual fiesta during the 1910s or 1920s. The building at the top of the hill was the Abousleman mercantile. The smaller building below it was the post office for some years. (Courtesy Abousleman family)

Mexico include parrots, copper bells, and traces of cacao (chocolate) inside specialized cylindrical jars.[8] Language studies also suggest connections via migrations of Uto-Aztecan into and out of the desert Southwest.[9]

Interestingly, Leslie A. White, an eminent ethnologist who worked at Zia Pueblo in the 1940s and 1950s, relayed the following story of Aztec-Pueblo contact from Fray Zárate Salmerón's 1626 *Relaciones* report. Zárate Salmerón was the Franciscan priest who, in the 1620s, directed construction of the Mission de San José (now Jemez Historic Site in Jemez Springs):

> It is possible that the pueblo country might have been visited by Indians from Mexico, Aztecs in particular, before the coming of the white man. Fray Geronimo de Zarate-Salmeron reported in his

Relaciones that a Spanish soldier had told him he had seen pictures of Aztecs, which he recognized by their dress, in a kiva at Acoma. The Aztecs had come from the west and had spent a few days at Acoma. Because the Acoma had never seen people like them, they painted their likenesses on kiva walls. When the strangers left they went toward Sia. All this took place, the Spanish captain was told, a few years prior to his visit. "With this information," said Zarate-Salmeron . . . "I made great research; and asking the chief-captain of the pueblo of Cia . . . and other elders, if they had information of those peoples . . . he said yes; that he very well remembered having seen them, and that some of them had been entertained in his house. That this was a few years before the Spaniards made a settlement in New Mexico; . . ." The strangers went on to Jemez, also, where they spent a few days."[10]

Contact and trade between Aztec and Pueblos certainly existed, but it is still unknown if there were specific migrations from New Mexico to Tenochtitlan. In any case, it seems to me that the mixing, melding, and re-creation of Montezuma myths among Pueblo, Spanish, Mexican, and Anglo people for all sorts of historical, cultural, and political reasons are emblematic of New Mexico, a multicultural landscape.

CHAPTER 7

A Visit to the Valles in 1886

J. K. Livingston and his family mounted horses and rode from Santa Fe into the Jemez Mountains in June 1886 on a vacation tourist trip. Fortunately for us, upon their return, the editors of the *Santa Fe New Mexican* prevailed upon Livingston to write an account of their Jemez adventures. Several of his descriptions deserve some explanation and modern context, which I provide following Livingston's article, included here in full:[1]

VISIT TO THE VALLES, JEMEZ SPRINGS, AND MONUMENT CAÑON, A VERITABLE LAND OF MODERN WONDERS; A REGION TRIBUTARY TO SANTA FE WHICH ECLIPSES THE "GARDEN OF THE GODS": TO THE EDITOR OF THE NEW MEXICAN:

Without any question the brightest jewels in the whole coronet of our territory's natural wonders are hidden within that mountain range bounding our western horizon, known as the Valles; and some day, possibly in the not far future, they will do for New Mexico what the Yosmite [sic] has done for California, or the Yellowstone for Montana.

The old Jemez Hot Springs is a resort for health and recreation not in the slightest degree overestimated; but it is merely at the portals of still greater wonders and more marvelous fountains of health. One mile above the main spring and the hotel, the rocky sides of the cañon contract to a width of about 150 feet; and here, for centuries has been carried on a mighty struggle for mastery between the tumbling and noisy current of the river, and the silent, mineral-laden springs; resulting in the upbuilding of that wonderful structure, the "Soda Dam." The

Chapter 7

springs have almost had the best of the fight for, in graceful draperies of creamy white, they have "built better than they knew;" and, with ramparts as wide as the bottom of the cañon, and forty feet in height, have turned the river fully 100 feet from its original course, and tumbled it in a very undignified but picturesque manner over a precipice, some thirty feet, where within the cool shades of a roomy grotto it sputters and boils, and then settles down into a respectable river again. Here, by virtue of that sovereignty which belongs to squatters, the writer, for nearly three weeks held with family and friends undisputed title to a tented patch of ground, with the walls of the dam in the immediate background, the road and spray of the cataract on one side, and over all a narrow strip of blue sky crowded.

BETWEEN TITAN WALLS of red and white and gray, rising abruptly for 2,500 feet; whose tops were fringed with lofty pines, which needed more than the sense of vision to convince the reason that they were not scrubby piñons.

It is an absolute fact, that within one hundred feet of camp there were not less than forty-seven distinct hot springs, and that each member of the party, even to the 2-years old baby, had a private bath tub, chiseled by nature in the rocks, and filled at all times with sparkling soda or sulphur waters, of just the right temperature for bathing; that we frequently took natural vapor baths, by visiting a certain grotto on the north side of the dam, within the walls of which we also enjoyed copious morning draughts of a most delicious hot mineral water, very similar in action, if not in taste, to that of Saratoga's most popular spring, the "Hawthorn," where the waters of the latter are heated artificially, and so furnished visitors when desired; and lastly,

Figure 7.1A–B. (*opposite*) A view of Soda Dam in 1910, by Jesse Nusbaum, compared to a view in 2014. Notice the old road ascending the dam, just visible at the far left in the 1910 photo, and the road cut in 2014. The increased size of the cascade domes over the river and the higher riverbed level are also noticeable in 2014. (1910 photo courtesy New Mexico History Museum, Palace of the Governors Photo Archives, NMHM/DCA Negative No. 012806)

that twelve feet above our canvass domicile, and easily reached by steps cut out of the rock, we fitted up a library and reception room, or rather, nature fitted it up, in the solid rock; of dimensions thirty feet long, ten wide and thirteen feet high; with an entrance twelve feet wide, and a window (overlooking the cataract) three by four feet, whose embellishment of halls ceiling and pillared portico, were in pure bi-carbonite of soda, with some adulteration of lime and alum—after the fashion of all baking powders, if we are to believe certain rival advertisements!

For the benefit of those of your readers who have previously visited this spot, it may be well to say that the unprecedented freshet in the Jemez river, this past spring, which rendered necessary the rebuilding of nearly the whole length of the cañon road, also, so we were informed by Col. Perea and others, completely changed, the location of nearly all the springs, from the top of the high rocks, to the edge and level of the river. But we are still at the threshold, and the wonders increase as we advance up the canon.

The wagon road ends at the sawmill just above the Dam; but taking saddle horses we rode up

TO KELLY'S RANCH, a distance of twelve miles, under the guidance of Kelly jr., aged 11 years, who justly earned the soubriquet of "Peck's Bad Boy," by the relation en route of startling and original jokes, in which bears, catamounts and senior members of the Kelly family were jumbled up in the most approved style.

The trail follows the west fork of the river, and is steep in places; diving into glades of pines and white birch, with vistas of cool waters tumbling under a wealth of woodbine arches, and anon leading far up the rocky sides of the cañon, with wonderful abutments of stone above and the feathery tops of pines hundreds of feet below. It certainly would not be an exaggeration to say that we passed springs every quarter of a mile, on an average, during the ride up the cañon; some of them iron but mostly soda or sulphur; and at one point, dismounting and visiting the river's edge, the inevitable pools of hot water were found, with vapor coming out of the crevices in the rocks.

Kelly's is an ideal stopping place for the weary mountain traveler, who will find within the friendly shelter of his log cabin rest, health and good cheer.

In all directions, the sloping sides of the basin in which these varied wonders are locked, are covered with sulphur refuse, from which we selected large nuggets of beautiful crystals, and over which one is inclined to tread very gingerly, as it is an uncanny spot with half hidden suggestions at every step of subterranean boilers and furnaces. Before breaking camp we made an excursion to "Monument Cañon," and "Valles de los Indios," which, for scenic grandeur far exceeded anything we had yet beheld. Leaving the cañon road at the ruins of the old Jemez church, mounted upon burros and with a Mexican guide, the ladies especially, viewed the undertaking with some misgivings, and well they might; for within a distance of two miles actual travel, the almost perpendicular side of the cañon was to be scaled to a height of 2,500 feet.

The trip, both going and coming, however, was safely made, and the view from the top was of course grand beyond description, as the BOTTOM OF THE CAÑON was at our feet, while looking south, 100 miles of the Rio Grande Valley lay spread out like a map.

A ride of three miles through a dense pine forest, brought us to the edge of Monument Cañon into whose depths we gazed with awe, as upon a Titanic Necropolis; a spot not belonging to this world, brooded over by a strange, dead silence, which affected the senses like an opiate. The cañon was same 200 feet deep; about the same distance in width, and possibly a quarter of a mile in length; the sides being absolutely perpendicular.

Into this yawning gulf were crowded some 150 monuments, turrets and spires of all sizes and shapes, varying in height from 40 to 150 feet, and each surrounded by large boulders of conglomerate rock, the material from which the structures were fashioned being a gravelly clay of cement-like hardness.

The cañon calls to mind the "Garden of the Gods" at Maniton, Col., but so far exceeds it in every respect as to make a comparison seem absurd. The highest structure which we christened "Trinity Steeple," greatly resembled in shape and height that famous New York landmark; and it was flanked by innumerable smaller spires of possible 100 feet in height, as closely crowded together as are the pinnacles on the roof of the great cathedral at Milan.

Large pine trees, scattered here and there among the monuments, gave a greatful [sic] touch of color to the strange scene, and perhaps added to its appearance as a "city of the dead."

Those who visit Jemez without seeing Monument Cañon miss one of the greatest scenic attractions in the territory; and yet, we are told, that comparatively few who visit the springs think of going there.

J. K. L

Soda Dam: Livingston's count of "not less than forty-seven distinct hot springs" on and around Soda Dam is similar to one of the earliest descriptions of this remarkable geological phenomenon by US Army explorers during their trip into the Jemez as part of the Wheeler Expedition in 1875. They counted forty-two springs, including twenty-two located on the dam itself.[2]

Today there is only a handful of significantly flowing springs on and around Soda Dam, including a couple that emerge on the west side of NM 4, where the road was (unfortunately) blasted through the dam in 1960 or 1961. The reasons for the decreased number of flowing springs are not clear, but it could have something to do with changing climate or maybe an increased extent and density of forests in the Jemez in the twentieth and twenty-first centuries; the forests take up a larger proportion of meteorological water. (See chapters 22 to 25.)

One of Livingston's most intriguing descriptions is of a rather large cave within Soda Dam with "dimensions thirty feet long, ten wide and thirteen feet high; with an entrance twelve feet wide, and a window (overlooking the cataract) three by four feet." There are no visible caves within or beneath Soda Dam today matching this description. (See chapter 25.)

Figure 7.2A–2B. (*opposite*) These photographs of Monument Canyon were taken in 1890 and 1998. This place is relatively unknown today because it is located near a little-used forest road, and forest growth since the late 1800s has obscured most of the hoodoos and monuments. (1890 photo courtesy New Mexico History Museum, Palace of the Governors Photo Archives, NMHM/DCA Negative No. 147691; 1998 photo courtesy Craig Allen)

There are many historical photos of Soda Dam extending back to the 1880s, but almost all are views of the south side (downriver side) of the dam. Photos of the old road that went up and over the west end of the dam are also rare. (See the 1903 photos of a freight wagon ascending the road on the south side in chapter 24.)

One last note about Livingston's Soda Dam description: His statement and quote that the springs have "built better than they knew" is most likely from Ralph Waldo Emerson's poem "The Problem." That poem includes these lines: "He builded better than he knew;— / The conscious stone to beauty grew."

Kelly's Ranch: This is most likely the hotel and hot spring establishment that was located at Sulphur Springs, just northeast of today's La Cueva. This was a well-known and popular destination for visitors and tourists to the Jemez Mountains for many decades. Old-timers in the valley may even remember taking baths at Sulphur Springs as late as the 1960s before the hotel and bathhouse burned down. (See chapter 34.)

Monument Canyon: This is a nearly forgotten place, despite being one of the earliest mapped, scenic places in the Jemez. The Wheeler Survey party made note of Monument Canyon as they traveled by it in 1875 and included it on one of the earliest maps of the Jemez published in the report.[3] An early photograph of Monument Canyon suggests why it was so named (see the old and new images). Perhaps the Wheeler surveyors thought the hoodoos resembled the many statues, obelisks, and other monuments erected in that era upon pedestals, commemorating heroes of the Civil War.

The Padre Alonzo Trail was an ancient travel route between Santa Fe and San Diego Canyon. This is the trail that Livingston and his party rode up on burros from the old mission. (See map of Monument Canyon and the trail in chapter 1.) The trail can still be found where it ascends through the rimrock of Cat Mesa above Church Canyon and several other places as it travels to the north end of Cat Mesa and passes by Monument Canyon and then down into Vallecitos de los Indios.

CHAPTER 8

John Wesley Powell's Jemez Dam Dreams

Droughts and floods are a fact of life in New Mexico. Given the vast amounts of snowmelt flowing down our rivers during spring some years, it is apparent that our problems are water excess at times and scarcity at other times. The idea of capturing water behind dams during wet seasons and releasing it during dry seasons was first championed for the arid lands of the western states by Major John Wesley Powell, the second director of the United States Geological Survey (USGS).[1]

Powell lost most of his right arm during the Battle of Shiloh in 1862, when he was an artillery officer in the Union Army. Seven years later, he became famous after leading a daring expedition down the uncharted Colorado River through the Grand Canyon. Over time, his 1869 adventure created a heroic reputation, highlighted by the story of the one-armed Powell running the Colorado rapids, sitting in a chair strapped to the deck of a small wooden boat.

Powell was a visionary explorer, scientist, and government agency leader. He began the first organized anthropological studies of North American Indian tribes. As USGS director, starting in 1881, he organized geographers, geologists, and surveyors to inventory and map the country's rivers, landscapes, and minerals. More than anyone else at the time, he understood that water availability would limit the development of the western United States. Powell seized upon the idea of planning and building dams and irrigation canals to solve the problem of seasonal aridity and multiyear droughts.[2]

A little-known historical fact is that Powell envisioned the Jemez Mountains becoming a national exemplar of converting desert to farmland using the Valles Caldera as a vast water storage tank. The *valles* and the canyons to the south were to become lakes held back by six dams. A lengthy canal would deliver stored water from the lowest lake in the Jemez Valley to the west mesa of Albuquerque, where, according to Powell, extensive farmlands would flourish.

Powell described this grandiose plan to Congress in February 1889.[3] The following are excerpts from a transcription of his oral testimony to the House of Representatives:

> A branch of the Rio Grande [is] a stream called the Jemez. I chose the Jemez River because I had previously made a geologic study of the river ... [and other studies] during the fall and winter in order that I might exhibit to the committees of Congress what this work would be when accomplished. On that river the surveys are not yet completed.
>
> There is the Jemez River [pointing to a map] which is a tributary to the Rio Grande del Norte. Here is Bernalillo. Here's Albuquerque. [At] about that line [illustrating on the map] we have a little system of mountains all over there we have a great section of mountains.... The whole following one great system and forming one great group of volcanic mountains. The region a little above that line—except a little portion next to the river—all of that region there is too high for agriculture. There is great rainfall on these mountains.
>
> Here we have great towering volcanic mountains, and the water gathers there and forms the Jemez River, and on reaching Jemez Pueblo it enters the lowlands and becomes broad and sandy. The stream, which above is a clear, narrow torrent, [becomes] spread out into a broad sheet and runs through the sands. Further than that, along the lower valley, 4 or 5 miles wide, the winds drift the sands and carry them constantly in sand dunes, so there is always sand filling the channel of the river. Whenever the water gets down here it is absorbed very shortly.
>
> At high water time it rolls into the Rio Grande del Norte, and at low water it sometimes does not reach to this point, and sometimes not even to San Ysidro. This is an old country, and so far as I can learn, irrigation has been practiced here since 1710 ... and about 2,700 acres have been irrigated there.
>
> ... We surveyed the river, and found the water can be stored at these points [indicating on map]. This comes out at Jemez Pueblo, above the point there [indicating], where a dam can be constructed at

Figure 8.1. John Wesley Powell, circa 1891, near Flagstaff, Arizona. (Courtesy National Park Service)

little cost, that can divert the water into a canal and carry it there. This is a hilly bit of country through which the canal is to be taken, but by carrying the water out here with a canal 25 miles long, we can come on the plateau above Albuquerque and irrigate all this arid land. We find that there is water enough, if it is stored and carried out in that way, and not allowed to be lost in the sands below, to irrigate 155,000 acres of land. So that now, where they irrigate 2,700 acres of land, it is possible, by storing the water and constructing the works, which I shall describe, to irrigate 155,000 acres of land.

It is proposed to construct a reservoir here [indicating on the map]. There is a little settlement; four or five men have gone in there and raised potatoes, and some seasons they raised oats and barley and wheat; and some seasons they can not raise anything, because it is too

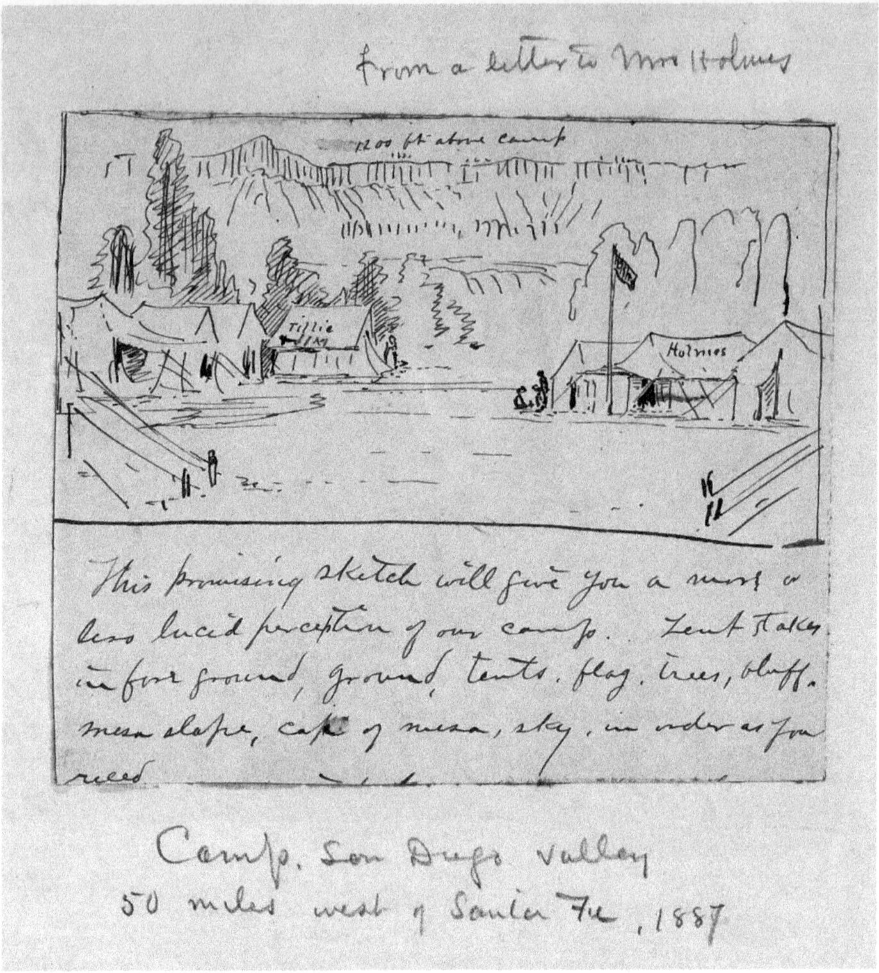

Figure 8.2. This sketch of the 1887 Powell, Stevenson, and Holmes expedition campsite is included in William Henry Holmes's memoirs. It was part of a letter home to his wife in Washington, DC. The tent labeled "Tillie & Jim" was Matilda and James Stevenson's tent. Holmes's tent is also labeled. The handwritten caption below the sketch reads, "This promising sketch will give you a more or less lucid perception of our camp. Tent stakes in foreground, tents, flag, trees, bluff, mesa slope, cap of mesa, sky, in order as you read." In addition to Powell, the Stevensons, and Holmes, the party also included the recently appointed secretary of the Smithsonian Institution, Samuel P. Langley. (From Holmes, *Random Records*)

cold; but there is a little tract of land up there which they cultivate, a few acres. The reservoir which is proposed to be constructed here (Valle Grande) will have a dam 500 feet long and 50 feet high, just here, below a mountain valley, so the whole reservoir will contain 63,000 acre-feet of water. . . .

    Now, with these six reservoirs we can store nearly all the water. I am speaking of these reservoirs now that will irrigate these 155,000 acres and redeem it. Now the water stored in these reservoirs has to be discharged during the irrigating season into the natural channels. It is only held back up in the mountain meadows, which are converted into lakes, and when the season for irrigation comes the gates of these reservoirs are to be opened and it is discharged . . . and to an extent that will be found true all over the United States, flowing waters must be stored high in the mountains and discharged during the irrigating season into the natural channels. The water so discharged is diverted by a dam here [indicating on the map] and carried by this canal.

    Powell went on to describe the locations of the dams and the total water capacity of the six reservoirs (155,125 acre-feet). Unfortunately, I have been unable to find the map to which Powell was referring in his testimony. However, from his descriptions, the dams would have flooded the Valle Grande, the Valle San Antonio, and the canyons of the Rio Cebolla, Rio Guadalupe, and Rio Jemez. The dam for the Valle Grande was probably planned for a spot just west of what is now the Las Conchas picnic site, where the East Fork first crosses under NM 4.

    Fortunately, Powell's plans for dam building in the Jemez fell apart when he faced increasing political pushback from Congress and territorial politicians. However, his general ideas about building dams for water control and irrigation became a reality in many places across the country. The Jemez Canyon Dam, which was ultimately built in 1953 near Santa Ana Pueblo, was designed for flood control and sediment retention rather than for irrigation purposes.

    After he was appointed director of the USGS in 1881, Powell worked primarily in Washington, DC, but he traveled regularly to the West, spending most summers camping and riding horseback in the mountains where he was planning dam sites. On these trips, he also pursued his interests in the ethnology of Native Americans.

He made at least a couple of trips into the Jemez Mountains. One of those trips, in August and September 1887, is described in the memoirs of William Henry Holmes, a member of that expedition.[4] Holmes was one of the first archaeologists to study ancient Pueblo ruins, and during the 1887 expedition he visited and mapped many of the largest ancestral Jemez village sites.[5]

The Powell expedition of 1887 camped in wall tents at a spot just south of what is now Hummingbird Music Camp. Holmes tells several interesting stories about the two-month-long campout with Powell and a group of eminent scientists from Washington, DC.[6] I include one of those stories, about grizzly bears in the Jemez, in chapter 15. Holmes had a close encounter with an *Ursus arctos horribilis nelsoni* (the Mexican grizzly bear) near their campsite.

Holmes also tells about of an ill-fated horseback trip he took with Powell to the top of Redondo Peak. They rode horses from their campsite to the summit, with Powell riding a large sturdy horse and Holmes on a smaller horse with a stiff-legged gait. Along the way, they stopped often to observe and collect archaeological and geological specimens, including artifacts from the sacred Hemish shrine at the summit.

Undoubtedly, Powell looked out over the landscape spreading below Redondo Peak and imagined lakes there filling the grassy valles. On the way down the mountain, the constant jolting of Holmes's stiff-legged horse apparently injured his back. The next day he could not walk and was in serious pain. James Stevenson, Powell's right-hand man, rigged up a litter attached to a mule, and they hauled Holmes down to Bernalillo for medical care. He recovered back in Washington, DC.[7]

CHAPTER 9

Dry Goods, Saloons, and Sheep

One of the most famous cultural landmarks in Jemez Springs is Los Ojos Bar. Established in 1947 by the Abousleman family, Los Ojos has been a place of so many "happenings" that it would take a book to describe even a small fraction of them. Here are a few brief stories about the early history of the Abousleman family in Jemez Springs leading up to the opening of the bar more than seventy-five years ago.

Moses Abousleman, born in 1869, arrived in New Mexico from Syria in the late 1880s and began a partnership in a dry goods store with another Syrian immigrant named Nathan Salmon. A wave of Syrian and Lebanese people immigrated to New Mexico from the 1880s to the early 1900s, and these families prospered in mercantile businesses and sheep herding. They were Christian Catholics coming from a not-too-dissimilar semiarid landscape.[1] The Salmon-Abousleman dry goods business, located on San Francisco Street in Santa Fe, was booming by the 1890s. In 1897 they decided to open a store in Jemez Springs in a building leased from the Otero family.[2]

There were plenty of customers for dry goods and other store products in the Jemez Valley at that time. A regular stage line ran from Bernalillo, and hotels in the village did business with tourists coming to soak in the hot springs, fish for trout, and for other recreation.

Shepherds and their large bands of sheep—numbering in the tens of thousands—moved up and down the mountain every year. In spring they headed up to the high pastures on the mesas and into the valles. In late fall they moved down to lower elevations. Abousleman was the primary partner traveling back and forth from Santa Fe to Jemez Springs with the task of setting up and operating the store while also establishing a sheep ranch.

In about 1900, Moses moved his wife, Edna, and their two small children from Santa Fe to Jemez Hot Springs. He purchased land and began

Figure 9.1. Advertisement in the *Santa Fe New Mexican*, June 5, 1900.

construction of the large, two-story stone house that sits today across NM 4 from Los Ojos Bar. Six more children were born between 1904 and 1918.[3]

Abousleman and Salmon amicably ended their partnership in 1904, and Abousleman received the Jemez Springs operations as his share of the business.[4] With his dry goods and sheep operations established, he expanded his ventures to a saloon, a hot springs bathhouse, and co-ownership of a stage line that delivered visitors to Jemez Springs from Bernalillo.

The Abousleman bathhouse was in the old "Hot Sulphur Water Baths" building, now in ruins, located near the current-day Jemez Hot Springs, formerly known as Giggling Springs. I have tree-ring dated several timbers and boards from that old building, which indicate a construction date of about 1895.[5] Abousleman bought the property in 1905 from L. A. Judt, who had run it as a bathhouse for paying tourists some years before.[6] Abousleman improved it with new tubs and paint in the 1920s. A 1941 flood ruined much of the building with mud, and it was used only intermittently by the family after that.

The early years of running the store and saloon were difficult, not least because the buildings caught on fire at least twice[7] and were also robbed multiple times![8] I'll return to the store robberies later.

By about 1910 Abousleman owned one of the largest sheep herds in the Jemez, with about forty-thousand head. The November 25, 1912, *Santa Fe New Mexican* described a gun battle between Abousleman and sheep rustlers:[9]

TWO DEAD IN MIDNIGHT DUEL

Moses Abousleman May Lose Arm

Terrible indeed must have been the duel fought at Zia Pueblo, Sandoval County, late Saturday night as the moon stood silver in mid-heaven while the reports of rapid firing rifles and pistols announced the clash between a band of five sheep thieves and a posse of nine law enforcers, including the owner of the 300 sheep which led to the trouble. The Dead: Baulito Lucero deputy sheriff of Sandoval County and Manuel Archuleta one of the alleged thieves [In a later correction, Archuleta was identified as a deputy sheriff, not one of the thieves.]

Figure 9.2. The original Abousleman store in Jemez Springs. Several different general stores and cafés were later built on this site. (Courtesy Adams family)

Figure 9.3A–9.3B. (*opposite*) An old postcard showing the Abousleman Hot Sulphur Water Baths and a modern view of the site.

The only one known to be wounded is Moses Abousleman, prosperous merchant and sheepman living at Jemez Pueblo and formerly partner of Nathan Salmon of this city. From the meager details which have been received here today the duel was between a gang of four or five sheep thieves and Mr. Abousleman and the deputy sheriffs aiding him in rounding up the robbers. It appears that Mr. Abousleman was informed Saturday evening while he was at his home at Jemez [Springs] that several prowlers had been seen around his ranch at Zia, 15 miles South of Jemez [Springs]. [Later reports said the fight was near San Ysidro.]

Losing no time, the merchant and deputies made a dash for the sheep ranch surprising the robbers at work. Then a terrific fight ensued, a deputy being shot down right after he had made one of the robbers bite the dust. Two bullets struck Mr. Abousleman in the arm and one

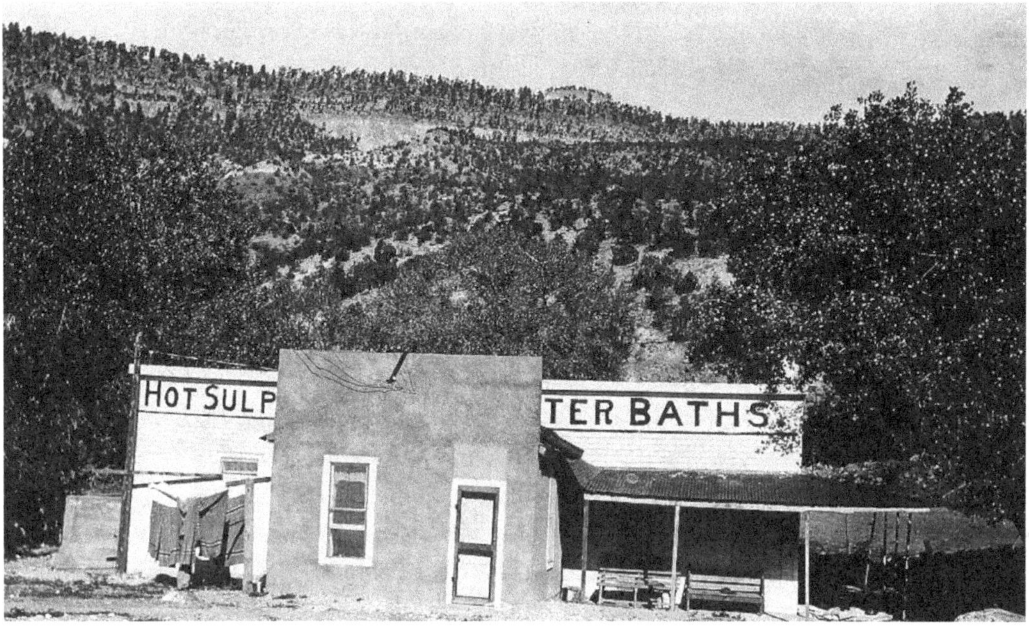
HOT SULP
TER BATHS

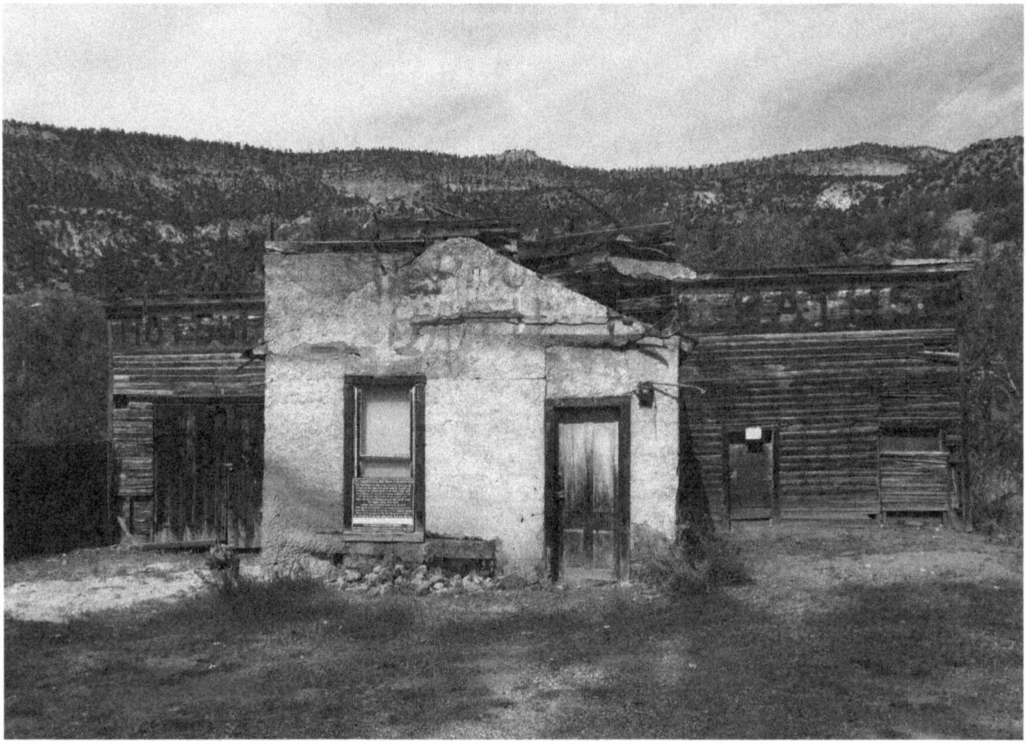

of the bullets shattered the elbow. Mr. Salmon was informed of the affair last evening and at once set out in his car for Jemez, with the intention of taking his former partner to a hospital in Albuquerque or bringing him here in order that the best of medical attention may be given him. Mr. Abousleman was reported to be in terrible agony as a result of his injuries.... The thieves escaped into the mountains where they are being pursued by a strong posse.

Despite being told his arm had to be amputated or he would die, Abousleman "steadfastly refused." By May 1913 he told his friend Nathan Salmon in a letter that his arm was gradually improving and he hoped to regain full use of it,[10] but he never did. I have been unable to learn whether the sheep thieves were ever caught.

Store thieves also plagued the Abouslemans, with at least three robberies reported in the newspapers between 1903 and 1931. The last event was probably the most dramatic, as noted in the article that follows. This robbery occurred in the same building where Los Ojos Bar is today. According to the March 18, 1931, *Albuquerque Journal*:[11]

ROBBERS CAUGHT AT JEMEZ SPRINGS

Sandoval County officers armed with sawed-off shotguns surprised and captured three men who had entered the general merchandise store of Moses Abousleman at Jemez Springs late Tuesday night.... The three men James Cox, Sigfredo Casados, and Epifanio Montoya all believed to be from the Jemez region, were surrounded in the store just after they had gained admittance with a skeleton key at 10:00 o'clock. Each was carrying a revolver.

Sheriff Montoya and two deputies Harold Golleher and former sheriff Mariano Montoya had come from Bernalillo and lain in wait inside the store with Fred Abousleman, son of the owner, since 7:00 o'clock in the evening. The officers posted themselves behind an office railing and young Abousleman was hidden near a string that had been tied to an electric light. When the door opened stealthily, he jerked the light on and the sheriff's men jumped up and covered the intruders with shotguns.

"Don't make a move or you'll get hurt!" warned sheriff Montoya. Cox a man of 40 years or more threw up his arms and sank to his knees begging for mercy according to the officers. Casados also elevated his hands at once, but Epifanio Montoya youngest of the three showed fight. A youth still in his teens he cocked his revolver to fire but changed his mind when prodded in the back by a shotgun in the hands of Mariano Montoya.

Casados who was about 30 years old collapsed after the capture just as the officers were starting back to Bernalillo in automobiles. He is believed to have suffered a heart attack and was placed under medical care at the county jail in Bernalillo. Both Cox and Casados are married, but the latter is said to have several children. Young Montoya, unmarried, told the officers he has spent his life in the Jemez Mountains and has never been as far away as Bernalillo and Albuquerque.

Mr. Abousleman became suspicious last Saturday when he saw Casados surreptitiously surveying a burglar alarm wire that runs from the top of the store to the Abousleman home across the street. Later, Casados tampered with the wire and a test by the proprietor showed that it was out of commission. He and friends kept watch in the store Saturday, Sunday, and Monday, and finally called officers from Bernalillo when learning that the holdup would be attempted Tuesday night. Two other confederates were believed associated in the enterprise according to Sheriff Montoya, but Cox when questioned about them said "I am no stool pigeon." Casados was unable to be interrogated because of his condition, and Montoya was evasive in answers.

Of course, all of this took place before the Abousleman mercantile building was converted into Los Ojos Bar in 1947. Some of the antique guns and other artifacts on the walls of the old bar are from the Abouslemans' Wild West days of the 1890s to the 1940s, but there are relatively few now compared to before about 1970, when the family sold the place.

CHAPTER 10

First Automobile in Jemez Springs

Residents of the Jemez Valley come to know quite well the curves and straightaways, the hills and vistas, as one drives NM 4 down the valley to Jemez Pueblo and San Ysidro, and then along US 550 to Bernalillo. The next time you drive this stretch, imagine what it was like before there was a paved road and, before that, not much of any road.

By the early 1880s, at least one set of ephemeral wagon wheel tracks seemed to exist. The new hotels in Jemez Springs began advertising in the Albuquerque newspapers, and by 1894 a "Star Route" postal service delivered mail twice weekly, when weather and roads permitted. Visitors describing the trip often mentioned the difficulty of crossing the sandy areas between Bernalillo and San Ysidro.

The following 1882 account of the trip from Bernalillo to Jemez Pueblo is from the diary of the intrepid twenty-five-year-old missionary Mary Lodisa Stright. She had traveled from Pennsylvania to join her former teacher Isabelle Shields and her husband, the Reverend John Shields, at the Presbyterian mission school and church.[1]

> Wednesday Oct. 18, 1882. After dinner, a Mexican young man came to the hotel and said he was going to Jemes in a buggy and if I wanted to do so could go with him. I did not know what to do, I was so tired of staying, but he was a stranger. I asked the landlady if she knew this young man and if it would be all right for me to go with him. She said he was a nice young man and would take me there all right....
>
> It would have been better if I had waited. Went about a mile or two when we came to the Rio Grande River which we crossed on a very long bridge and then we were into the sand which was at least 6

inches deep. The first part of the way the plain was pretty well covered with clumps of cedar wood, some cactus and different kinds of weeds such as I had never seen before and would only see in such a place. The wonder to me was that anything at all could grow.

After awhile it got worse. The clumps of cedar wood were farther apart till at last nothing could be seen but great drifts of sand. Once in a while we caught sight of the Jemes River which runs through here. The Mexican said the road or tracks rather never stayed in one place very long, for in a wind storm, a common thing, those drifts might be carried over into the track.

His old horse is very pokey and he has to whip him to make him go at any kind of decent rate. I feel sorry for the horse but if we only get to Jemes I don't care. It begins to get dark and I ask him if we are nearly there, and he answers, that we are not halfway. I begin to feel a little suspicious. It is moonlight and we ride on. At last the horse will go no farther. We are near an Indian pueblo and the Mexican thinks he must stop to feed his horse. . . .

[Stright and her guide ate some food and then slept overnight in someone's home at Zia Pueblo.]

October 19, 1882. Morning came at last and, oh how thankful I was. . . . We soon started on our road again which was the same as yesterday for five miles, when we came to a Mexican town [San Ysidro] where I saw a large flock of sheep, more than a thousand. . . . At last we get to Jemes and stop at Dr. Shields, door. They seemed very glad to see me.

Tourists, government surveyors, foresters, and others continued to visit Jemez Springs in increasing numbers through the 1910s, conveyed only by horses, wagons, and stagecoaches. Finally, in 1912 the first automobile arrived, as described in this newspaper article:[2]

CASE CAR MAKES TRIP TO JEMEZ AND RETURN

Otto Berger and Party First Ones Ever to Negotiate Road to Springs in Automobile: Fast Time and No Mishaps

Following the route of the United States mail stage to Jemez Springs, Otto Berger of this city accompanied by George Hottinger as chauffer, A. J. Burkhead and J. S. Lovern, established a new record for automobiles yesterday when they returned from a trip to the springs in a Case "Thirty," owned by Mr. Berger. The sixty-five miles to the springs was negotiated easily both going and returning, though previously a trip to Jemez in an automobile was looked upon by motor car owners as an impossible feat. Fast time was made despite heavy roads and not a mishap of any kind occurred to interrupt the trip. The feat speaks volumes for the Case car.

The party started yesterday on a trip to Sandoval county with no idea of visiting Jemez Springs, according to Mr. Berger, but after negotiating the sands with ease, it was decided to head for the springs just to see how far the auto could go, knowing that no other auto party had ever ventured on such a trip. The car bowled along at a good clip and almost before those in the machine realized it the Case glided into Jemez.

George Hottinger, who drove the car, and who has had years of experience in sands and mountainous country was astonished at the way the car behaved on the trip. He says the time of the journey could easily have been cut down considerably had any attempt been made to establish a record for fast running. The party left Albuquerque at 12:53 p.m., reaching the springs at 6:04 p.m. the same evening. On the return trip they left Jemez yesterday at 7:03 a.m. and got into Albuquerque at 11:35, using chains on the tires because of mud. It is the opinion of Mr. Berger and his driver that the wide built under frame of the Case car made the run easy if not possible.

Figure 10.1. Magazine advertisement for a 1913 Case 30 automobile. The J. I. Case Company was most famous, both before and after 1913, for its agricultural and construction machinery.

Mr. Berger, who is the Albuquerque agent for the Case car, took pains to have the feat properly recorded so that it could not be questioned, by getting a letter from John Woodgate, proprietor of the Stone hotel at Jemez Springs, certifying to the trip made by the Case car. This statement follows:

Jemez Springs, N.M., Jan. 18, 1912

> The above party stayed at my hotel last night, this being the first automobile to come to this part of the country. It created quite an interest among the people as it was considered impossible for an automobile to come over these roads.

Yours truly,
JOHN WOODGATE, Proprietor Stone Hotel, Jemez Springs, NM.

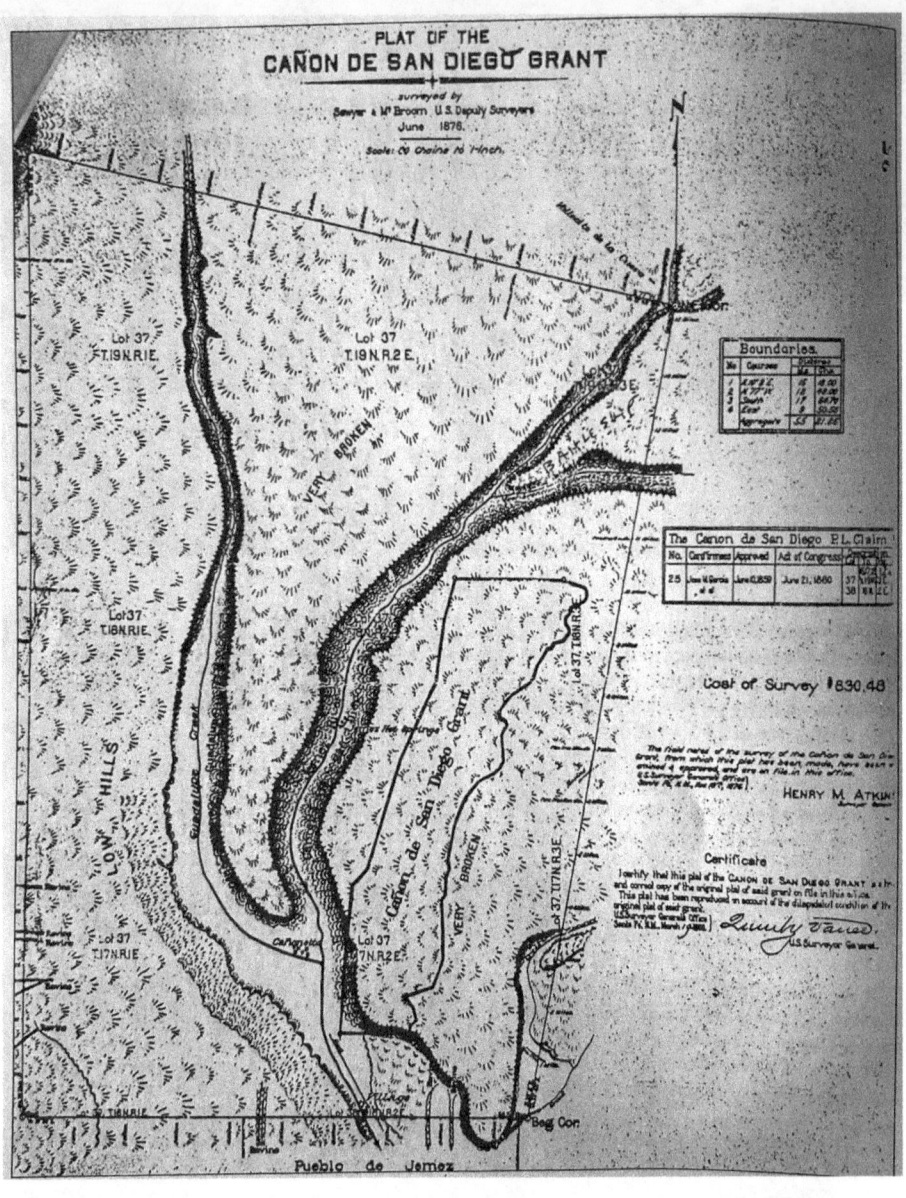

Figure 11.1. Cañon de San Diego Grant plat map, 1876. (Map by Sawyer and McBroom, deputy land surveyors with the US General Land Office)

CHAPTER II

Whose Lands Are the Jemez Mountains?

In March 2023 the US Court of Appeals for the Tenth Circuit issued a momentous ruling. It overturned a district judge's earlier decision and found that Jemez Pueblo retains "aboriginal title" to the Banco Bonito area of the Valles Caldera National Preserve.[1]

This is the southwestern corner of the preserve, including the "pretty bench" south of Redondo Peak. This rolling landscape created by lava flows more than forty thousand years ago is today covered in ponderosa pine forest and is crossed by NM 4.

The legal case is complicated and hinges upon how the courts have interpreted laws requiring "long time" and "exclusive use" of lands by tribes to establish that they still hold aboriginal title. The lands were assumed to be vacant by early US government surveyors and were granted in 1860 to Luis Maria Cabeza de Baca and heirs. The court determined that the act of Congress establishing Baca Location No. 1 did not include an explicit "extinguishment" of aboriginal title for areas of long time and exclusive use by the Hemish (Jemez) people.

It is unclear how this ruling will play out, as there are still many unknowns. Jemez Pueblo did not receive all the lands it claimed, leaving the potential for further legal action. Either the pueblo or the US government could appeal the case to the Supreme Court. If the Tenth Circuit Court ruling is appealed, it will not be the first time a legal case involving a land grant in the Jemez Mountains is decided by the highest court. A previous case involving the Cañon de San Diego Grant was also a complex series of legal decisions with many twists and turns throughout its lengthy history.

Disputes over the San Diego Grant began in the early 1800s because two overlapping land grants had been issued by authority of the king of

Spain to colonists in 1788 and 1798. Both land grants included many ancestral Hemish village and farm sites that had been depopulated since the late 1600s. Multiple litigations in Spanish and Mexican courts, and later in US courts, attempted to sort out which land grant should be recognized. The later 1798 San Diego Grant was affirmed in 1860 in US courts, but additional challenges were finally resolved by the Supreme Court in 1897 and 1899.[2]

The New Mexico territorial government passed a law in 1870 allowing the division and sale of Spanish community land grants, which was not allowed under Spanish and Mexican laws.[3] The 1870 law did not apply to the Pueblo land grants. Then, after the Supreme Court decisions of the 1890s settled its status, the San Diego Grant became an easy target for lawyers and speculators following the playbook of the "Santa Fe Ring."[4]

They were a group of powerful politicians, lawyers, and businessmen who controlled the politics and economy of New Mexico in the late nineteenth and early twentieth centuries. They used their connections to manipulate the legal system, making it difficult for Hispanic landowners to assert their rights and protect their land from encroachment. Their goal was to partition and buy community land grants at very low prices and then sell them to wealthy investors for huge profits.[5]

Using such schemes, a lawyer and his partners acquired the community portion of the San Diego Grant very cheaply from the Spanish American heirs. An interesting document that summarizes this history of what amounted to a legal swindle appears to have been written by John Amos Adams Sr. (or possibly his son John A. Adams Jr.).

The elder Adams was an early Forest Service man who around 1912 worked on a survey of the boundaries of the Jemez National Forest (which later became part of the Santa Fe National Forest). In 1914 he married Helen Shields, the daughter of the Reverend John Shields, who established the Presbyterian church in Jemez Springs in 1881.

The Adams document "Canyon of San Diego Grant"—a portion of which can be viewed on the Jemez Valley History web pages—relates that in the 1870s and later, Mariano S. Otero purchased multiple farm tracts along the Jemez River.[6] Each purchase included the transfer of rights to use of the surrounding community land grant. Otero, a wealthy businessman and politician, owned most of the Baca Location No. 1 lands by the late 1890s. The community portion of the San Diego Grant was about 110,000 acres,

Chapter 11

and the family-owned farm tracts along the Jemez and Guadalupe Rivers were around 6,000 acres.

Otero began using the community lands extensively for sheep grazing and logging. After he died in 1904, his heirs asserted that they had rights to use all the community land grant resources. This did not sit well with the heirs of the land grant in Jemez Springs and Cañon. The Adams document continues the historical narrative in the excerpt below:

> An enterprising young Albuquerque lawyer named A.B. McMillan [Alonzo B. McMillen was his full name and the correct spelling] evidently saw a golden opportunity in this situation, and he forthwith sent an assistant named Amado Chavez to explore the possibilities. Chavez reported that the heirs not having cash for lawyer's fees were glad to offer Mr. McMillan half of their interest in the grant if McMillan would fight the Oteros in court. McMillan agreed and Chavez secured the signatures of the majority of the heirs on a petition asking for a division of the grant among all the resident heirs and the Otero family, and naming McMillan as their representative.
>
> Mariano Otero had died in 1904 and his heirs appear to have lacked the aggressiveness of the old man. McMillan, a friend of the court, had an easy time proving the validity of the claims of his clients. The court found that the Oteros could prove ownership in only about 20% of the grant. McMillan's fee was one half of the 80% amount belonging to his clients, the resident owners.
>
> The court then named a Commission of three to determine the best method of dividing the grant among all the heirs. The Commission appears to have included several friends of McMillan.
>
> The Commission decided that because of the difficulty of equitably dividing such a heterogeneous area, the 110,000 acre commonly owned grant was to be sold in its entirety, and the proceeds prorated among the heirs on the basis of their proved ownership. The District Court accepted this proposal and in 1907 or 1908, decreed that the grant was to be sold at public auction.
>
> McMillan who still represented the owners, and indeed now owned approximately 40% of the grant himself, bid $0.45 per acre [$49,500 total cost]. As he was the only bidder, he got the grant. It

is stated in the area that Mr. Joshua Reynolds, [Joshua S. Raynolds was the correct spelling] of the First National Bank of Albuquerque, furnished the cash needed by McMillan, for the latter was still a poor young man. After the costs were deducted the heirs to the Canyon de San Diego grant finally received not more than $25,000 for it according to Mr. J.B. Block, who had bought an interest in the grant before the sale. [Block was a longtime hotelier in Jemez Springs.]

McMillan quickly found a prospective buyer of the grant he now owned alone in the White Pine Lumber company. A cruise of the timber showed that there were 165,000,000 board feet of merchantable timber on the grant, and the White Pine Lumber company indicated a desire to purchase if a way of taking out the timber could be found. Mr. McMillan secured a right of way for a railroad to be built into the grant from the owners of the lands in Jemez Canyon by pointing out to the owners that this would make wage work available near home.

The White Pine Lumber company bought the grant in 1910 for a price reported to have been more than $400,000. This company, which became the New Mexico Lumber and Timber company in 1920, held the grant as a reserve supply of timber until 1922. When the railroad to Camp Porter was finished in 1922, lumbering operations began on an extensive scale, and have continued at varying degrees of intensity to the present.

To sum up, lawyer McMillen, who was supposed to represent the interests of the land grant heirs, was the sole bidder at a very low forty-five cents per acre. He bought the whole 110,000 acres for himself and his business partners, and for his fee, he received half of the 80 percent of the proceeds due to the heirs!

Such a blatant conflict of interest and self-dealing would surely be grounds for disbarment and civil lawsuits in the modern era. But this was territorial New Mexico, and for the Santa Fe Ring it was business as usual. McMillen and his business partners got the whole community land grant for a song and then a few years later sold it to a logging company for at least a 1,300 percent return on their investment.

Although the Adams document suggests that McMillen was poor and new to this game, he was not. In fact, a few years earlier the team of

McMillen, Chaves, and Raynolds had followed a similar playbook of deception and self-dealing to reap a large sum in orchestrating the partition and sale of the Las Trampas Grant in the Sangre de Cristo Mountains. In that case, for his lawyer fee, McMillen received one-quarter of the proceeds due to the land grant heirs.[7]

What about the heirs of the Cañon de San Diego Grant? Well, hundreds of heirs were identified at the time of the partition and sale, so their typical payout was less than one hundred dollars each. Some of these heirs did not live in the valley or on the old land grant. Many could not read or speak English. Few, if any, had notification of the sale or understanding at the time of how their patrimony was being swindled by unscrupulous men aided by "friends in high places."

The former Cañon de San Diego Grant community lands, which were sold and then heavily logged starting in the 1920s, were eventually procured by the US Forest Service in the mid-1960s.[8] Today, they comprise part of the Jemez Ranger District on the Santa Fe National Forest. They are once again "community lands," but the "community" now encompasses all citizens of the United States.

There are small Hemish house ruins dating to the early 1600s and tree stumps from 1880s logging on the land where my wife and I reside and walk daily in Area 3. These historical remnants serve as a reminder to us that although this land is ours today in a legal sense, the Hemish people and Spanish land grant heirs once held title. And it was taken away unjustly—twice. At the very least, this reminder calls for humility and recognition of those who came before us, their descendants, and their ancestral lands.

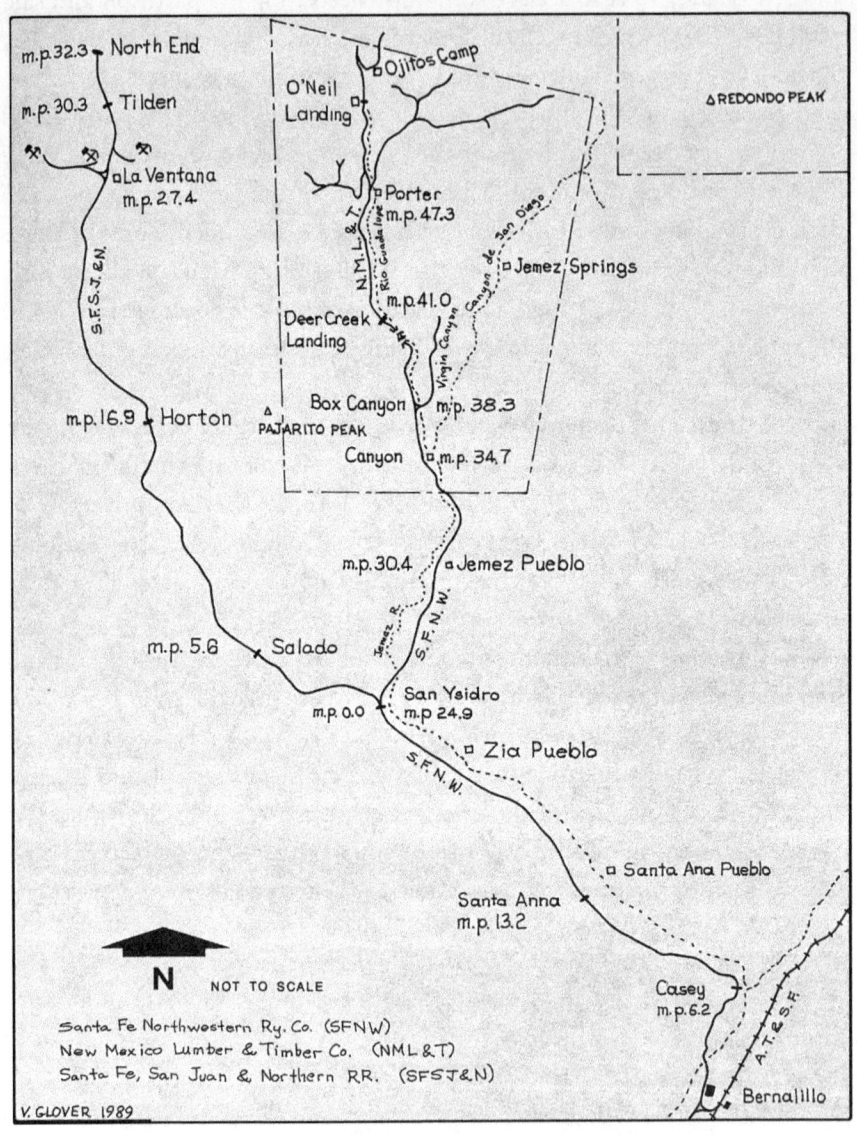

Figure 12.1. Vernon Glover's map of the Santa Fe Northwestern Railroad in the Jemez Mountains. The abbreviation *m.p.* stands for "mile post." The longer dashed lines mark the boundary of the Cañon de San Diego Grant. (From Glover, *Jemez Mountains Railroads*)

CHAPTER 12

The Santa Fe Northwestern Railroad

The arrival of the Atchison, Topeka, and Santa Fe Railway (AT&SF) in New Mexico in 1880 transformed the economy, culture, and landscape of the territory. One of the developments was increased tourism promoted by AT&SF, especially the visitation of hotels at scenic places with hot springs. The four-hundred-room, Queen Anne–style Montezuma Hotel was built in Las Vegas, New Mexico, in 1881. Thousands of visitors came by rail to soak in the hot springs and to hunt and fish in the Sangre de Cristo Mountains. Guests during the first decades of operation included Theodore Roosevelt, Rutherford B. Hayes, Ulysses S. Grant, William Tecumseh Sherman, John C. Frémont, and Jesse James.[1]

The idea of a similar railroad and hotel development at Jemez Hot Springs captured the imagination of Mariano S. Otero, who already owned land in the Jemez Valley and later most of the Baca Location. Otero was an insider in Santa Fe politics and business, so he was well aware of AT&SF's ambitions in New Mexico. In anticipation of a tourist boom, he and other investors began building a hotel at the hot springs in San Diego Canyon. The Stone Hotel, with about a dozen rooms for rent, was completed in 1882. Otero's partner and brother-in-law, Colonel Francisco Perea, moved with his large family to Jemez Hot Springs to manage the hotel and bathhouse.

Otero, Perea, and other investors lobbied AT&SF to build a rail line, and in 1882 the New Mexican Railroad, a subsidiary of AT&SF, developed plans for a line from Bernalillo to Jemez Hot Springs. However, for various reasons the dream of passenger trains delivering tourists to the Jemez never materialized. It wasn't until after the Cañon de San Diego Grant was swindled away from the original owners in 1908 (see chapter 11) that a railroad line to the Jemez became a reality. But its purpose was hauling timber, not tourists.[2]

The valuable timber on the San Diego Grant was primarily located in the canyons and on the mesas to the west of Jemez Hot Springs. To access the timber, a railroad was planned from Bernalillo to San Ysidro, up the Jemez River through Pueblo lands, and finally up the Rio Guadalupe and its tributaries. Civic and business leaders of Bernalillo donated land for a new lumber mill located on the east bank of the Rio Grande just north of town.

The Santa Fe Northwestern Railroad (SFNW) was incorporated in 1920. Its original aims were to build a rail line to La Ventana near present-day Cuba and to use the line for hauling coal to Bernalillo. The timber potential of the old San Diego Grant, however, took precedence. In 1922 the timber rights on the grant were sold by its lawyer and banker owners to the White Pine Lumber Company (WPL). According to Vernon Glover, "The purchase price was set at $2.00 per thousand board feet, the amount to be determined by a later timber cruise. Later the parties agreed to conclude the transaction on the estimated basis of 400 million board feet." That is, the estimated selling price for the timber on the old land grant was $800,000 in 1922 dollars, equivalent to about $14.6 million today.[3]

Construction of the railroad to the upper Jemez Valley first required building a long trestle across the Rio Grande and then laying the railroad bed across the sandy desert and many arroyos along the route from Bernalillo to San Ysidro. The most time-consuming hurdle, though, was obtaining the right-of-way through Jemez Pueblo lands. Hemish historian Joe Sando tells this sad story of another land rip-off in his book *Nee Hemish*. The railroad engineers planned to build the railroad next to the river through some of the best farmlands of Walatowa. Tribal leaders, however, refused to agree to this.[4]

Initially the US Department of Interior permitted the construction through the Pueblo lands, overriding the tribe's objections, but legal challenges continued while the railroad was built. Ultimately, in 1926, the powerful railroad and timber interests managed to get New Mexico's US senator Holm O. Bursum to introduce and push through Congress a far-reaching bill called the Pueblo Lands Condemnation Act (revised in 1928). This act enabled the federal government to "condemn" Pueblo reservation lands, allowing the development of public rights-of-way for roads, railroads, and electric power lines. So Jemez Pueblo was forced to live with the already-built railroad through some of its best farmlands, with the low compensation of $2.50 per acre.[5]

Figure 12.2. The big trestle just below the Guadalupe Box under construction, circa 1924. In 1939 it burned down and then was quickly rebuilt. (From Glover, *Jemez Mountains Railroads*)

Over the following fifty years, Pueblo tribes throughout the Southwest lost thousands of acres of lands to public right-of-way developments invoking the Condemnation Act. This sweeping injustice, not applied to any other tribal lands in the United States, wasn't rectified until 1975, when the act was repealed. The repeal allowed some pueblos to recover part of their lost lands but not all. Subsequent acts of Congress provided additional compensation.

In 1923, after several years of construction and with ongoing arguments over the right-of-way through Jemez Pueblo, the railroad reached the Guadalupe Box. This was by far the greatest engineering challenge to the railroad builders: how to get through the narrow granite gorge to the main timberlands in the canyons and mesas above it?

Undaunted, they built a massive trestle over a tributary canyon just south of the box and then blasted two tunnels through the bedrock. As

Figure 12.3. Early log truck in the Jemez, circa 1932. The trucks got the job done, but safety was another matter. (From Glover, *Jemez Mountains Railroads*, with credit to T. P. Gallagher Jr.)

Glover says in his history of the SFNW, "Although it was a spectacular and expensive piece of construction, the Guadalupe Box railroad was one of the least known and least remarked upon mountain railroads in the southwest. Three-eighths of a mile in length, this segment of the SFNW cost about $500,000 to build, more than half the cost of the entire railroad."

Despite many delays, including floods in early 1924 that washed out parts of the rail line, logging commenced in the summer of 1924. By the end of that year, total expenditures by the SFNW on railroad construction and equipment were $971,336. The WPL was now employing two hundred men in the woods above the Guadalupe Box and one hundred men in the mill at Bernalillo. The influx of railroad workers and loggers created a population boom in the Jemez Mountains. The 1920 US Census for Precinct 10 of Jemez Springs tallied 172 people. This precinct included the Jemez Springs, La Cueva, and Cebolla Creek areas.[6] The 1930 US Census listed a total of 677 people, a 390% increase since 1920! Many descendants of the families that

Figure 12.4. A loaded Mack truck with a wide log bunk in one of the Guadalupe Box tunnels, around 1950. The tunnels had to be enlarged to fit these big trucks and their loads. (From Glover, *Jemez Mountains Railroads*, with credit to the Yale Weinstein Collection)

came to the Jemez to work the rails and woods remained in the Jemez, even after the railroad was gone and most logging ceased.

The railroad logging of the upper Guadalupe and tributaries continued, with many ups and downs during the years 1925 to 1940. Fluctuating timber prices, the Great Depression, a labor strike, floods, burned trestles, derailments, and a locomotive boiler explosion all caused slowdowns and temporary work stoppages in the woods and at the mill. The ownership of the timber rights and company leadership also changed hands multiple times. The WPL and the SFNW were bought by the New Mexico Lumber and Timber Company (NML&T) in the early 1930s.

Chapter 12

The Hemish people of Walatowa reluctantly tolerated the railroad tracks and trains passing through their farmlands, but there were tensions and conflicts. Joe Sando describes how train engineers began a habit of stopping their trains in the fields during late summer to pilfer watermelons. The farmers retaliated by placing logs on the tracks under a train car while the engineers were busy stealing fruit. After they started up again the logs "jolt[ed] the car off the tracks, melon eaters and all."[17]

By the late 1930s, most of the timber easily accessible via railroad was cut out. From the beginning, log chutes and horse teams had been used to skid logs to landings, where they were loaded onto rail cars. By the mid-1930s, gasoline-powered trucks hauled logs down dirt roads to the landings and, later, all the way to the sawmills. At about this time, NML&T expanded its truck logging operations to the Baca Location, where it acquired timber rights.

Decreased profits and mounting railway expenses, plus the alternative of truck logging, meant the railroad days of the Jemez were numbered. Then, in spring 1941, a huge flood ripped out roads and bridges all along the Jemez River and its tributaries (see chapter 27). Several miles of the railroad and multiple trestles were also washed out. NML&T decided not to rebuild. Steel prices also went way up after the start of World War II, so the rails were pulled up and sold for scrap.

With abandonment of the rail line, Jemez Pueblo regained its farmlands. In a later controversy over the right-of-way, the state wanted to build NM 4 along the old railroad bed through the farmlands, but this time

Figure 12.5A–B. (*opposite above*) View up the Rio Guadalupe above the box, with Joaquin Mesa on the horizon. The 1904 Forest Service photo shows the old growth along the canyon bottom and slopes before the logging. The 2021 image shows the second-growth forest with fewer tall old trees. (1904 photo from Benedict and Reynolds, *Proposed Jemez Forest Reserve*; 2021 photo courtesy Steve Sutherland)

Figure 12.6A–B. (*opposite below*) View just south of Guadalupe Box showing the trestle that burned in 1939 and was quickly rebuilt. Burned logs and boards can be seen on the slope above the river in the 1944 photograph. Willows have grown up over the river in the 2023 photo. (1944 photo courtesy Jemez Ranger District, US Forest Service)

the Pueblo succeeded in thwarting that plan. The compromise was locating the roadbed where it is today, on the far eastern edge of Walatowa, along the base of the low sedimentary terraces there.

In 1948 a sawmill and tepee burner were built just below the Guadalupe Box at the Gilman landing. For decades this was a main hub of logging and milling in the Jemez Mountains. In 1965 the US Forest Service finally acquired the old San Diego Grant lands. Truck logging by NML&T continued along the upper Guadalupe and tributaries until 1973. After that time, logging continued on the Baca Location and elsewhere on the Santa Fe National Forest.

From the 1930s until at least the 1980s, logging trucks of various sizes were a very common sight, sound, and hazard along roads and highways in the Jemez. Bruce Lamb, one of the assistant rangers during the 1965 transfer of the San Diego Grant lands to the Forest Service, recalled driving what is now Forest Road 376 through the tunnels and beyond. The big logging trucks used along that road segment had extra-wide log bunks, so two vehicles could not pass along the road. The truck drivers used CB radios to let each other know where they were, so one of the trucks could stop in a pullout and wait for the other truck to pass. The Forest Service trucks didn't have the same radio bands, so Bruce would open both windows and anxiously hang his head out so he could hear if one of the big trucks was coming down the canyon. If so, he'd pull off anywhere he could or begin to back up fast!

CHAPTER 13

The Jemez National Forest

In 1904 the New Mexico Territorial Legislature and Governor Miguel A. Otero II objected to the proposed establishment of the Jemez Forest Reserve, arguing that it would drive away private landowners, inhibit mining and timber development, and limit other uses. However, President Theodore Roosevelt ignored their protests and established the reserve in 1905. That same year, he created the US Forest Service with Gifford Pinchot as its first chief. Conservation and wise multiple uses of resources on publicly owned lands were set as prime directives for the new agency. The reserve was huge, including large parts of the Jemez Mountains and extending to lands north of the Chama River all the way to the Colorado border.[1]

In 1907 the reserve was formally designated the Jemez National Forest. The portion north of the Chama was split off and added to the Carson National Forest in 1908. In 1915 the southern Jemez portion was combined with the Santa Fe National Forest. The eight-year-old Jemez National Forest was gone, but the Forest Service was here to stay.[2]

Over the following decades, forest rangers made their presence felt, riding horseback dozens of miles every day, talking with ranchers and loggers, building fences and trails, setting up grazing allotments and permit systems, marking timber, fighting fires, and doing many other tasks. They built ranger stations, guard stations, and lookout towers, and they strung telephone lines between them. Some early rangers packed sidearms because the task of regulating grazing and timber cutting was sometimes contentious. But almost always, they used diplomacy rather than force.[3]

One of the first Forest Service men to develop a long-term connection to the Jemez was John Amos Adams. He started his career in 1909 on national forests in Arizona. In 1912 he was assigned to the Jemez National Forest with the job of mapping the new forest.[4] The early rangers who surveyed and mapped the original forest reserves were referred to as "forest arrangers" by Gifford Pinchot. They were instrumental in laying out boundaries and locating

Figure 13.1A–B. The original ranger's residence in Jemez Springs, built in 1929. Perhaps the men in the photo are Perl Charles and Jim Curry. The second photograph is the one-room ranger station office. Behind the office you can see a corner of the old warehouse and tack room. There was a small corral behind that building. (Courtesy Southwestern Region, US Forest Service)

sites for future lookouts and ranger stations. While in Jemez Springs, Adams met Helen Belle Shields, the daughter of Reverend John Shields, the medical missionary who in 1881 founded the Presbyterian church in the village. They were married in Albuquerque in 1914. Descendants of the Adams-Shields family have owned property and lived in Jemez Springs since then.

The Jemez Ranger District of the Santa Fe National Forest did not become a separate administrative unit until the 1920s. The large San Diego Land Grant and the Baca Location split the forest lands in the Jemez into isolated parcels east-west and north-south of these private landholdings. The main ranger stations were located in Española, Cuba, Coyote, and the mining town of Bland on the Pajarito Plateau. Guard stations were built at strategic locations, including at Las Conchas, La Cueva, Seven Springs, Rio de Las Vacas, and Blue Bird Mesa. These were typically one- to three-room log cabins where the rangers could sleep overnight during bad weather and their horses could be held in a corral. Sadly, all these old cabins and related structures are gone now, demolished in the 1970s or later as a cost-cutting measure or burned up in wildfires.

The administrative center of the Jemez River Ranger District (the original name), located in Jemez Springs, began in 1929 when the first ranger station and residence were constructed at the mouth of Church Canyon, across the arroyo from the old mission church ruins. The first district ranger here was Perl Charles. He was an experienced ranger, having previously served on the Lincoln and then the Santa Fe National Forest, working out of Cuba. You can get a sense of the independence, self-sufficiency, and toughness of these old-time rangers from the following excerpts of an interview with Charles in the 1960s:[5]

> In the spring of '24, I moved to Española and worked under Earl Moore for a few months and then took charge of what they called the Santa Clara Ranger District at Española. Most of that District, I believe, is now in the Los Alamos Atomic Energy area.... In 1929 we moved to Cuba, and we lived there for a few months. They were building a new Ranger Station at Jemez Springs. We moved to Cuba in March, I believe, and the Supervisor, Frank Andrews (he was a wonderful man) said, "We'll have that new Ranger Station ready about the first of July and you can move down there." The first of July came along, and I asked him about it, and he said, "We didn't get the money we thought we

would. We have the walls up on the Ranger Station, and we have part of the roof on it, and we have the plumbing in. If you want a place to live, you and your assistant will have to finish it." So, we did. Later on, we had some complaints about the carpenter work, but it was about the first carpenter work I ever did. Jim Curry was the assistant. He and I had to finish the Ranger Station because they ran out of money. That was a very interesting District. I stayed there until 1935.

There was a fellow named Dick Wetherill who later became my assistant, or Administrative Guard, as they called them then. He was a natural woodsman; he spoke Spanish as well as he did English, and Navajo just as well.... The big business on that District was grazing. We took care of fires on the Baca Location, and we had to work with fires on the San Diego Grant because they would burn toward us.

The ranger's residence in Jemez Springs, which was finished by Perl Charles and Jim Curry in 1929, still stands on the old Forest Service compound next to Jemez Historic Site. The original ranger's office was in a one-room building. It was torn down after a new ranger station was built in 1966, when my father, Fred Swetnam, was the district ranger. Our family lived in the old residence from 1964 to 1973.

Dick Wetherill, who was mentioned by Perl Charles, worked at the Jemez District for at least two decades. One of his primary duties was keeping the ground-return, hand-cranked telephones operating. This required constant maintenance of many miles of single-strand #9 wire telephone lines. He was the son of the famous Richard Wetherill Sr., an amateur archaeologist who partially excavated the cliff dwellings at Mesa Verde and later Pueblo Bonito at Chaco Canyon. Wetherill Sr. was murdered by a Navajo man at Chaco Canyon in 1910.

Perl Charles's account of his troubles in dealing with livestock issues on the national forest shows how challenging and dangerous a ranger's work could be in those days:[6]

When they finally came up with this system under which we could gather [trespass] horses, or round them up, and finally got the Forest boundaries fenced ... we started in on the horses. We took over 1,500 head of horses off that one District.... The only problem we really had with the stockmen was—there were just a few of them then—was the horses, those trespass horses, they belonged to a very few men.... We had some pretty close shaves with those fellows. There was nothing we could do but try to talk 'em out of it, and personally, I had a lot of experience with that. I couldn't run very fast, and I wasn't big enough to fight, so I had to talk 'em out of it.

A fellow by the name of Joe Rodriguez was the Ranger on the Coyote District. He was a wonderful man. His father-in-law was an ornery old cuss. He was sent to the pen for killing a man, and he finally was killed by a Deputy Sheriff. I remember we had a bunch of horses up there in Rock Creek and he went in there into the corral. I said, "What are you doin' in there, Manuel?" He said, "I'm gonna get my horses," I said, "You can't get 'em," and he said, "I am." So I just walked out in the middle of the corral there with him. There were a couple of dozen Coyote natives and they all had their weapons of one sort or another. There were about three or four of us—there was nothin' for me to do but talk them out of it. I just stood out there in the middle of the corral and argued with that guy, musta been two hours. Finally, I said, "You can take your horses, the horses you want, and they'll be $3.00 apiece," or whatever the cost was. He said, "I don't have three dollars." I said, "I'll give you a job and you can work it out." And that's the way we settled it. I gave him a job in camp; he didn't do anything, but he stayed around long enough to work it out.

We had a lot more freedom in those days. You could collect the money right there and usually we could pay those fellows. You could pay your own expenses. Now of course everything has to go through channels. Life was much more simple then.

In 1935 Perl Charles was transferred to other work in the Santa Fe National Forest, and Leonard Ward Lewis became the new Jemez District ranger. He was also an experienced ranger, having worked on the Coconino and Manzano National Forests. He usually worked every day except Sunday, and sometimes then too. In the summer he often traveled more than two hundred miles a week via a combination of horse, car, truck, and foot. All rangers were required to keep a daily diary of their work. The following are entries from Lewis's 1937 diary:[7]

The Jemez Mountains

Date: Monday Feb. 22, 1937;
Mileage: by horse, 10; Expenses: $1.05

8:00AM: Left by horse & rode up Church Canyon & up telephone line almost to main Hughes road. Cut back south to Forest Boundary.

Saw signs of 2 small bunches of horses. Arrived Hughes mill at 2:00 pm & marked timber until 4:30 pm.

5:00PM: Stayed at Hughes Mill tonight.

Date: Tuesday July 27, 1937;
Mileage: by car, 12; by horse, 40; Expenses: $1.40

8:00AM: Packed up car from road camp & left for Seven Springs.

10:00AM: Caught horse & rode up Hay Canyon & to boundary fence & inspected.

1:00PM: Rode Sulphurs S&G [sheep and goats] allotment.

4:00PM: Counted Francisco Gurule's sheep & told them to move to Jarosa allotment Aug. 20. Saw six head of Routledge horses & 30 cows & stock on Sulphurs S&G allotment.

6:00PM: Started down Barley Canyon & ended up in Bear Canyon (lost, raining, dark). Rode across to road camp in East Road Fork.

10:00PM: Arrived [Seven Springs]

Date: Thursday Sept. 9, 1937;
Mileage: by car, 20; by horse, 10; Expenses: $1.05

8:00AM: Left Station & to Abousleman mill by car. Saw Lew Caldwell & talked concerning sawdust & slab disposal & sanitation.

1:00PM: Went to Caldwell Ranch & talked to Mrs. Caldwell concerning G. [grazing] C&H [cattle and horses] permit. Rode from Caldwell's to Grant line & thru some of timber cut near line. Not close enough to require a fire line. Rode along Grant boundary & found place where there is no fence crossing East Fork; on across by Garcia & Maestas place; across under Los Griegos & to Maestas salt ground. No salt. Rode back around under Los Griegos saw several head of cattle, all permitted.

6:00PM: Returned to Caldwell's & stayed the night.

Figure 13.2. Leonard Ward Lewis in the 1930s, probably on the Coconino National Forest. (Courtesy Lewis family)

It is exhausting just to read Ranger Lewis's list of extensive travels and many tasks carried out every day across the sprawling Jemez Ranger District. Some of the entries in 1938 refer to mechanical troubles with his truck and difficulties in hauling his horse by trailer to various places where he would then unload and ride horseback for many miles.

Tragically, on November 1, 1938, his truck, pulling the horse trailer, slipped off the road and plunged into the canyon between Battleship Rock and La Cueva. Both he and the horse were killed. He left his wife, Hazel, and nine children.[8] Hazel and some of the children stayed and worked in the Jemez for decades, contributing in many positive ways to our community. Grandchildren and great-grandchildren descendants are still here today continuing that tradition.

Figure 13.3. This sign was posted on the district ranger's office door in the 1950s. The name of the ranger and the single day of the week the ranger would be in the office could be changed. This sign was from either El Rito or Peñasco ranger stations on the Carson National Forest or from the Capitan District on the Lincoln National Forest, where Fred Swetnam was the ranger in the 1950s before being transferred to the Jemez District.

Fred Swetnam was posted to his first ranger district, at El Rito, New Mexico, Carson National Forest, in 1952. My mother, Grace, and eldest brother, Jim, eighteen months old at the time, moved into an old adobe ranger's house there. My dad's daily workload at El Rito was much like those of Rangers Charles and Lewis, who spent perhaps one day a week in the office and five or six days on horseback, driving a truck, and walking the woods. By the time he and our family were transferred to Jemez Springs in 1964, he was much

Figure 13.4. The Seven Springs ranger station in about 1965. Fred Swetnam stands on the ground. The sign above the porch says, "Seven Springs Ranger Station." Old Forest Service maps label it as a guard station. Some of the early rangers and their families lived here during the summer months. (Courtesy Swetnam family)

more like the modern ranger who feels lucky to get out in the woods one day a week while working in the office on administrative matters other days.[9]

Appointed district rangers come and go on the Jemez District at about two- to ten-year intervals, but over time the longest-serving men and women are local Hispano and Pueblo people with deep roots here. These include the Sandoval, Waquie, Romero, Casaquito, Gachupin, and Gonzales families. The Sandovals have a long history of service on lookout towers and other district jobs. Pedro Sandoval and his son Simon, for example, manned the Cerro Pelado Lookout for most years from the 1910s to the 1950s, and his nephew Moises Sandoval manned the Bearhead Lookout in the 1920s. Moises continued working on the district for decades, and two more generations of Sandovals have worked careers here, including his son Gilbert and granddaughter Juanita.

The old San Diego Land Grant and Baca Location are now in public ownership. The third Jemez Ranger District office was completed and moved into in 2023. It sits pretty much in the same place where the 1966 office and warehouse were located. The new office has two floors and even includes an elevator. Perl Charles, Leonard Lewis, Moises Sandoval, Fred Swetnam, and all the early Forest Service men who knew the original one-room district office would be amazed!

Finally, a note of remembrance for eight men who lost their lives while working for the Forest Service on the Jemez Ranger District.

- Leonard W. Lewis, district ranger, 1938, truck accident
- Abenicio Juan Diego Toya, wildland firefighter from Jemez Pueblo, 1946, wildfire accident
- Andrew V. Waquie, Benjamin P. Waquie, Allen M. Baca, Anthony Pecos, wildland firefighters from Jemez Pueblo, 1986, truck accident on a wildfire in Idaho
- Frankie Toledo, wildland firefighter from Jemez Pueblo, 1993, wildfire accident
- Token Adams, wildland firefighter, 2013, all-terrain vehicle accident associated with a wildfire

CHAPTER 14

Peeled Ponderosa Pines

When you visit Valles Caldera National Preserve, be sure to see History Grove. This is an old growth stand of ponderosa pine and Douglas-fir located about a half mile northeast of the old ranch headquarters. The main road heading toward the other valles goes right through the grove. The former owners of the Baca Ranch, and others who held the timber rights, logged most of the old growth that once grew around the rims of the valles. Fortunately, the rancher landowners preserved most of the large, ancient trees in this grove, as well as other old growth around the ranch headquarters. Some of these trees are more than four hundred years old.

The Dunigan family owned the Baca Ranch before the US government purchased the property, and they called this beautiful spot Sherwood Forest. Sometime after the purchase, the name History Grove seemed to stick, possibly because some of the largest ponderosa pines there have peculiar scars on their boles that tell stories of people who were here long ago. Several of these trees can be seen adjacent to the road as it enters History Grove on the south side.

Resource managers call these culturally modified trees, or CMTs, and a particular type of CMT is sometimes called a peeled ponderosa pine. Probably the first written historical record of this kind of CMT in the western United States was by Meriwether Lewis and William Clark as they traveled through the Bitterroot Mountains of Montana on their way to the Pacific Ocean. Clark's journal entry on September 12, 1805, included this sentence: "I mad(e) camp at 8 on this roade & particularly on this Creek the Indians have pealed a number of Pine for the under bark which they eate at certain Seasons of the year, I am told in the Spring they make use of this bark." Later, on May 8, 1806, Lewis mentioned inner bark use in the context of discussing famine among the Shoshone people.[1]

The Lewis and Clark descriptions, and others over the past two centuries, often associate the eating of the soft tissues of the inner bark with hard times. However, the record is muddled, with a variety of other uses for the bark also reported. Some records suggest it was a seasonal delicacy, was used for making fermented drink, or was used for medicinal or other purposes.

I'll come back later in this chapter to the mystery of why Indigenous people of the southwestern United States peeled ponderosa pines, but first, here is the oldest mention of peeled trees in the Jemez, from a 1904 report by surveyors with the Bureau of Forestry (which preceded creation of the US Forest Service): "In most of the pine areas, girdled or partially girdled trees are seen occasionally. This is done by the Indians who scrape off the inner bark and use it as a food in times of unusual scarcity."[2]

There likely were many more peeled trees in the Jemez before the extensive logging era of the twentieth century. But they are still found as single trees or small groups of peeled ponderosa pines in the Valles Caldera and on the Jemez Ranger District. I am aware of about a half dozen locations with one to a dozen or so trees in each: La Cueva and the Horseshoe Springs area, Seven Springs, Calaveras Canyon, San Juan Mesa, History Grove, and old Baca Ranch headquarters area.

Peeled trees and other kinds of CMTs are especially interesting as "living artifacts" because they often can be studied using dendrochronology—that is, tree-ring dating—to determine when the modification took place. The dating is accomplished by taking core samples with an increment borer. This is a long hollow drill bit with a handle. It is used to extract pencil-size cores from the boles of trees that have been peeled. The method involves taking a core sample through the edge of the peeling scar and then crossdating the ring-width patterns with a master tree-ring width chronology for the region. Dendrochronologists use this procedure to match ring patterns that result from climate variations and hence are common among all trees in a site or region.[3]

Using an increment borer, I have cored and tree-ring dated about a dozen peeled trees in the Jemez. The peeling of the trees at History Grove took place between the 1810s and the 1830s. The trees near Calaveras Canyon (north of Seven Springs) were peeled in the 1810s. These dates are comparable to those of dozens of other trees that have been sampled and dated elsewhere in New Mexico. The peeling dates are usually in the historic period, from the late 1700s and to the late 1800s.[4]

Figure 14.1A–B. Peeled ponderosa pines at Calaveras Canyon. An increment borer was used to take cores through the healing curl around the two sides of the peel and from the back of the tree. By matching the ring width patterns among the different cores and with a master tree-ring width chronology from the Jemez, it was possible to determine the approximate date of the peeling. In this case, the date was about 1810.

Figure 14.2A–B. A different type of cultural modification: Christian crosses carved into two trees in History Grove, Valles Caldera National Preserve. Bill deBuys, the first chairman of the original Valles Caldera National Preserve Board of Trustees, stands next to one of the crosses in about 2001. The tree-ring date of that carving is approximately 1917 and was probably created by sheepherders.

Chapter 14

The practice of peeling trees for the inner bark is probably much older, but perhaps it became more frequent during the turbulent times of the eighteenth and nineteenth centuries, especially during campaigns by Spanish colonial and then US Army soldiers against Navajo, Ute, and Apache. There is relatively little ethnographic documentation of the use of inner bark by Pueblo people (except for the Zuni). However, it is mentioned more frequently in accounts from the early twentieth century and later studies of plant use by Great Plains and northern Rockies tribes and by Navajo, Ute, and Apache in the Southwest.[5]

The interpretation of an emergency food source during warfare was supported by the first tree-ring study of peeled trees I conducted, about forty years ago, focused on a group of trees in Lilley Park, Gila Wilderness, New Mexico. The peeling of these trees was done in the spring of 1865. Subsequent to my tree-ring dating of those trees, I found a probable historical connection to the peeling event: A band of Gila Apache, led by Mangus Coloradas, was pursued around southern New Mexico by the US Army under General James Carleton throughout 1862. The rugged and remote headwaters of the Gila River, now encompassed by the Gila Wilderness, were a favorite refuge of the embattled Gila Apache. Then, as described in this excerpt from my 1984 paper:

> Victorio became the leader of the Gila Apache after the murder of the captive Mangus Coloradas by his soldier guards in January 1863. Skirmishes between soldiers and the Gila Apache continued into 1865, when Victorio attempted to make contact with Carleton to discuss peace. Carleton sent an emissary in May of 1865 to talk to Victorio in the Gila area [at Pinos Altos], and he later quoted Victorio as saying . . . "I and my people want peace; we are tired of war; we are poor and we have little for ourselves and families to eat or wear; it is very cold; we want to make peace, a lasting peace. . . ." Unfortunately Carleton was intransigent, and lasting peace did not come until after the death of Victorio and most of his warriors in 1880 in a battle with the Mexican General Terrazas.
>
> . . . It seems quite clear that the Gila Apache were experiencing unusual hunger during 1865 and 1866. Although the historical evidence is circumstantial, it appears to be a very compelling argument for the emergency food utilization pattern, at least for the Gila Apache and the peeled trees in Lilley Park.[6]

Since my 1984 paper, there have been several other tree-ring studies of peeled trees in New Mexico and Arizona, including those I have done in the Jemez and most recently a study of a very large set of peeled trees near Picuris Pueblo.[7] There are more than three hundred peeled ponderosa pines at that location, which is the site of the 1854 Battle of Cieneguilla between Jicarilla Apache and US Army dragoons. The dragoons were nearly wiped out during that engagement.[8] The dating of peeled trees in these other locations by me and other researchers does not confirm the "hard times" hypothesis, but they don't clearly reject it either. In general, the peeling dates tend to be spread out over multiple years within sites, and although they sometimes coincide with wartime periods, they are not consistently associated with fighting or with droughts.

So we are left with a continuing mystery. It seems likely that ponderosa pine trees in the Jemez were originally peeled by Jicarilla Apache, Ute, or Navajo people. It is also possible that these living artifacts were created by Pueblo people, but I have not heard oral history or read documentary evidence of that for Jemez Mountains peeled trees at this time. Although the jury is still out, I think the hypothesis that use of inner bark was related to hard times remains compelling; that is, inner bark was used especially during times when people were experiencing some kind of duress. And it is possible that inner bark was used mainly for medicinal purposes during hard times rather than as an emergency food source.[9]

Perhaps one day this mystery will be solved, and maybe not. In any case, keep your eye out for these special trees and marvel at their historical and mysterious presence.

CHAPTER 15

Four Jemez Mountains Grizzly Bear Stories

If you have lived in or visited the Jemez for significant periods of time, you have probably seen a bear. The bears here today are black bears (*Ursus americanus*), and usually you see them running away from you. They can be dangerous, though, especially if you come between them and their cubs. This apparently happened some years ago on the Valles Caldera National Preserve during a marathon. The runner was badly injured by a mama bear, and following state law the bear was euthanized. The cubs were captured, raised in captivity, and released some months later.[1]

Grizzly bears (*Ursus arctos horribilis*), also called brown bears, used to populate the Jemez, and their presence created a particular sense of danger for those traveling through the woods. If you have ever hiked around in the northern Rockies in grizzly bear country—carrying pepper spray if you are wise—you know what I am talking about. Four stories that tell about grizzly encounters in the Jemez provide a glimpse of what it was like to share this landscape with those fierce creatures.

The first story is from 1887, when William Henry Holmes camped with Major John Wesley Powell and other famous scientists just below what is today Hummingbird Music Camp (see chapter 8). His grizzly bear encounter took place somewhere below the cliffs of Virgin Mesa, probably near what is now known locally as Area 3.[2]

> For a long way I rode up over an ancient village site, then up sharp ridges among the timber until I came to a flattish timbered shelf that lies along the base of the final ascent. Here at the elevation of about 1000 feet above camp I found many small ruins and some pottery. The final step of the plateau consists first of a steep slope up which I had

to lead my mule zig-zagging back and forth over the rocks and slides. This slope ends against the base of the capping cliff which is in the main nearly vertical and from 100 to 300 feet high. It extends so for many miles.

I hitched my mule on a little shelf at the base of this cliff and began to look for a place reduced or broken down sufficiently to let me climb it. As I skirted the base of the cliff to the right I happened to look down the steep slope below and there, about 20 feet below, was a grizzly bear. He was nosing along and did not see me, but he was going right toward my mule and I concluded very quickly that that would not do, since by going 20 feet further he would give my mule such a fright that he would break loose and rush down the mountain.

I had no gun or pistol so I shouted "Boo, hoo," at the bear. He glanced up quickly and saw me, and made a spring away from me, facing down the steep slope. At this moment I pushed off a big stone and sent it after him, flying.

The result was "too funny for anything." The mountain was very steep for a long distance below and covered with loose stones and scattering trees. Down this slope the bear plunged and the big stone and many other loosened stones after it-rattle, bang, crash-until the cliffs re-echoed the uproar. I never saw a beast make such time and the stones were more rapid than he and made enormous leaps until they caught up with him and both, with many added stones, went out of sight together down into a rocky gorge nearly half a mile below me.

It was a laughable termination of the incident, but a good riddance of an ugly customer. The wild mountain declivities echoed probably for the first time in their history with roars of amused laughter tinged possibly with a shade of relief on the laugher's part. I soon reached the top of the cliff by a very ticklish climb, pulling myself up by little notches in the rocks, and the gooseberry bushes that grow in the crevices. From the summit I had a broad view of the valley and the surrounding mountains, made a sketch, and cut my initials and the date in the rock that forms the extreme point of a projecting shelf of the plateau and then, on account of a thunderstorm which suddenly broke across the plateau I hurried down to my mule. In the rain I pulled the unwilling animal by main force down the steep mountain face.

Taking a different course from the ascent I encountered a cliff midway in the slope and had a hard time, going back again and taking another spur and getting into camp late, wet and tired. The boys were quite excited that a bear should be so near and wanted to go on a hunt.

The second story is from an 1898 newspaper article.[3] I assume this bear was a grizzly, given the aggressive response to being shot. Too bad it wasn't the sheriff who got mauled because he started the trouble! It is also amusing that a fish story ends this tale, despite the bear mauling:

RETURNED FROM JEMEZ

A Bear Story in Which Messrs Finical and Powars Figure

Judge J. W. Crumpaker, Sheriff T. S. Hubbell and Col W. F. Powars returned late yesterday afternoon from their hunting expedition in the Jemez Mountains. They stopped at the Jemez hot springs, and from there went trout fishing on the San Antonio, up in the vicinity of Hugh Murray's place.

The following thrilling story is told on Col. Powars and District Attorney Finical, who accompanied the party to the springs but will remain there until his wounds are healed.

On the trip going to the river, the sheriff spied a big bear snoozing quietly in one of the numerous little caves at the Soda dam, and turning to the party remarked "I will clip out a portion of old Bruin's left ear with a 45-caliber bullet," and he shot, the ball wounding the bear. The beast, maddened by his wound which was up in the left shoulder, made a break for the tourists.

The judge and sheriff whipped their horses out of danger, but the bear wreaked vengeance on Messrs. Finical and Powars, who were trailing behind and failed to get out of danger's way. Both gentlemen were painfully lacerated, especially Mr. Finical, about the hands, and their clothing somewhat torn before the bear was killed. . . .

The judge and sheriff carried off the honors as fisherman, the former yanking out a four pounder and the sheriff coming in a good second with a two pound trout.

The third story was told by Joseph E. Routledge, a longtime resident of the Jemez Mountains:[4]

In 1917–18, my father started a homestead in a little canyon (Medio Dia) near Bland, just below Los Alamos. He had a team, a good double-bitted axe, and a saw. He cut down trees and dragged them to a level spot and built a log cabin on it all by himself.

It took all summer, and he installed my mother and my older brother, along with some chickens and a milk cow. From the milk, we could make butter, which could be traded at the mining company store in Bland, for flour, sugar and lard. He had planted an open patch of ground in potatoes, and those would be harvested and also traded in Bland. He built a corral, shelter, and fenced lot for the milk cow. The skunks ate his chickens, so they were kept under the porch of the cabin, but bad luck loomed ahead.

One morning the cow was not to be found, and he thought she had wandered off. As my father looked around, he saw blood and drag marks on the ground and found huge footprints of a great bear. The bear had killed the cow and dragged her across the Media Dia Creek and uphill into some trees. The carcass was partially eaten—but no sign of the grizzly bear. No other animal had the power to drag a milk cow across a creek and up a hill. The tracks were there to see.

My father found himself on foot and unarmed near a grizzly bear's meal. He went back to the cabin, saddled a horse, picked up a rifle, and went back to the half-eaten cow. He found the bear's tracks leading away to the north. He followed them for a while, but they disappeared into the pine needles. He watched for the bear to return, but it never did. Everyone was sure that it had left New Mexico and was the last of the grizzlies in the Jemez Mountains.

Chapter 15

Last, the fourth story is a Forest Service report from 1940, indicating that the bear Routledge's father saw in 1918 was not the last one in the Jemez. David E. Brown, one of the top wildlife experts in the Southwest, wrote in his 1985 book, *The Grizzly in the Southwest: Documentary of an Extinction*,[5] that grizzlies were in decline by World War I but were still present in both Arizona and New Mexico into the early 1930s. Brown stated, "In 1917, Ligon [an early New Mexico wildlife biologist] estimated that only 48 grizzlies remained in NM—in the Sangre de Cristo, Jemez, San Mateo, and Magdalena Mountains; the Mount Taylor and Black Range; and the Mogollon Mountain complex." After the early 1930s, "Years passed without further verified accounts of grizzlies" in Arizona or New Mexico.

Brown continues, citing a 1940 US Forest Service wildlife report, "Several grizzly bears are reported on the Jemez District of the Santa Fe Forest. On several occasions, bear hunters have had their dogs badly torn up by bears that will not tree. Tracks and feeding habits of these bears along with hair from one animal that one dog got hold of prove the existence of grizzly in the area."

This was the last official report of grizzlies in New Mexico.

Part II

Stories from Richard Baxter Townshend

Tent rocks and pine trees below Virgin Mesa.

Figure 16.1. This plate is from James Simpson's report on his 1849 trip to Jemez Springs. The drawing is by expedition artist brothers R. H. and E. M. Kern, who visited the hot springs in the center of the village and the old mission church. (From Simpson, *Navaho Expedition*)

CHAPTER 16

Wild West Days of the Jemez Valley

The written history of the Jemez Valley begins just before the seventeenth century. Old Spanish documents recount a visit by Juan de Oñate to the valley in 1598 and the building of the Mission de San José at Gíusewa (Jemez Historic Site) in the early 1600s. There are also descriptions of the Hemish role in the Pueblo Revolt of 1680, and multiple battles that took place on and near Guadalupe Mesa during the reconquest of the 1690s.[1]

After resettlement of the Hemish at Walatowa (Jemez Pueblo) around 1700, there is almost no documentary history about the upper part of San Diego Canyon during the eighteenth century and only a few more sources on the early nineteenth century. The San Diego Land Grant was issued to about twenty Spanish families in 1798.[2] However, settlement on the grant was apparently minimal during the early 1800s, except for the lower portion near the junction of the Guadalupe and Jemez Rivers at Cañon (also called Cañoncito in some documents and early maps).[3]

From the earliest written accounts in the 1800s, it seems that "Los Ojos Caliente" (Jemez Springs) probably did not have permanent residents until sometime after 1850. In chapter 4 I quoted Lieutenant James Simpson regarding his 1849 observations of the Spanish Queen Mine, located a few miles south of the present-day village of Jemez Springs.

Simpson does not mention anyone living in upper San Diego Canyon then, but he says there were abandoned "old adobe buildings" along the river below the hot springs. He said the settlers deserted them because of attacks by Navajo raiders. Another story tells of Navajo killing two Jemez Springs residents sometime in the 1800s. (See chapter 19 about the killing of the Archuleta sons.)

Before the Navajo were defeated in 1864 by Colonel Kit Carson and the US Army, and then marched to an internment camp at Bosque Redondo, San Diego Canyon was a dangerous place to live and visit.[4] The reason for the delay of permanent settlement of Jemez Springs is also suggested in an 1857 account of a visit by a trader named Franz Huning. He later became a wealthy merchant in Albuquerque. In his memoirs he wrote,[5]

> It must have been about this time that I made my first trip to the Jemez Hot Springs, as my rheumatism had not entirely disappeared. I went on horseback alone. As there was no accommodation at the springs at that time, I had to take my blankets along, so that with other necessary things my horse was pretty well loaded down and therefor my progress was slow. On the first day I only went to Corrales.
>
> The next day I went to the Pueblo of Jemez where I camped with the priest. The next day early in the morning I reached the Cañon, which was a very lonesome place without a human being all the way up to the Springs. The only sign that anybody had ever been there was a cluster of two or three smelting furnaces with piles of copper ore and slag close by. I felt very lonesome in that wild cañon all by myself and was much relieved when I arrived at the Springs.
>
> It was lonesome enough even there, since a Mexican and his wife were the only people there. There were two log cabins, in one of which the Mexican couple rusticated, in the other there were two wash or bathings tubs, into which the hot water ran directly from the spring. The only comfort there was consisted of a few boards laid down alongside the bath tub, on which to spread whatever bedding a person might have brought along for sweating and sleeping . . .
>
> My landlord had cautioned me from the first against strolling away from the Springs too far, as the Navajos paid occasional visits to the neighborhood. I did go though on horseback as far as the waterfall

Figure 16.2A–B. (*opposite*) R. B. Townshend took the first photograph on his 1903 return trip to Jemez Springs. In one of his books, he mentions encountering Navajo riders, presumably these two men. (Courtesy Pitt Rivers Museum, University of Oxford, Catalog No. 1998.58.129)

several times. On one of these visits I met my old friend Manuel Abrego, as already stated, who was located at the springs farther up the creek, now called the Sulphurs. He invited me to come and visit him at his new rancho, but I did not go; I believe I was afraid of the Navajos.

The family that hosted Huning at the hot springs in 1857 was most likely José Francisco Archuleta and Maria Viviana Archuleta, who owned property in that part of the village, as documented in later decades.[6] The post office and the village were called Archuleta on official documents and an 1890 US government map. Many descendants of the Archuletas, including the Sandoval, Garcia, Montoya, Cart, and other families, live in the valley today.

The most colorful stories about the Jemez Valley during the late 1800s were written by Richard Baxter Townshend. He was a twenty-three-year-old British adventurer who came to southern Colorado in 1869 and started a ranch. He tells of those adventures in a book called *The Tenderfoot in Colorado*.[7] He sold the Colorado ranch in 1874 and he bought freight wagons and trading goods. Then, with two partners he traveled to the Jemez Valley, where they set up shop. It is not entirely clear where their trading post was located, but by his description it sounds like it was in Cañoncito.[8]

He quickly befriended John Miller, a government agent living at Jemez Pueblo. Miller later moved to Jemez Springs with his family and built a home near the old mission church. Townshend also befriended Emiterio Archuleta, a son of José and Maria living in Jemez Springs. Townshend tells several stories involving his friend "Miterio" in his books.

One partner who came with Townshend from Colorado was Vicente Elias. Their partnership ended when Elias was murdered at their trading post. The exact reason why he was shot dead through a window one night is unclear, but Townshend says it was apparently because Elias had "rashly allowed himself to take sides in certain local political factions, which were very bitter." This was indeed the Wild West![9]

Seeking justice, Townshend went to Bernalillo to talk with Don Francisco Perea, who he described as a "Mexican grandee." Perea told him, "Esperate . . . Wait, have patience! Their side is in now, and it is perfectly useless to bring them before any of the courts. The murderers of that side are protected by their bosses. But our time will come! Some day we shall

Figure 16.3. An R. B. Townshend photo from 1903, with this caption: "Jemez Indian with his burro loaded with apricots." (Courtesy Pitt Rivers Museum, University of Oxford, Catalog No. 1998.58.26)

win the elections and we shall be back in power again; and then, oh, but they will suffer for their sins!"[10]

Townshend does not indicate that justice was ever served, but Perea certainly was influential in those times. He had been a Union Army colonel at the Battle of Glorieta Pass in the Civil War, a territorial representative in Washington, DC, and a personal friend of Abraham Lincoln. He was present at Ford's Theatre when Lincoln was assassinated in 1865.[11]

Perea and his family moved to Jemez Springs in about 1880, along with the wealthy and politically connected Mariano Otero family. Perea's first wife was Dolores Otero, Mariano's sister. These close families built and operated the original stone-walled bathhouse, which survives today in the village center, and the 1881 Stone Hotel, which is now part of the Bodhi Manda compound.

After the murder of his partner, Townshend sold his Jemez trading business. He returned to England in 1877, later becoming a classics scholar at Oxford University. He spent his elderly years writing magazine articles about the American adventures of his youth. These stories were later compiled into the *Tenderfoot* books.

In addition to the book on his Colorado ranching days, he wrote *The Tenderfoot in New Mexico* and *The Last Memories of a Tenderfoot*. *Last Memories* includes stories of Townshend's nostalgic return visit to Jemez Springs in 1903. He stayed with his old friend John Miller at his big house across from the ruins of the old mission church. He also visited with Miterio Archuleta and hiked around Soda Dam and the mesas. In addition to the Elias murder story, Townshend's books include other fabulous stories of people and places in the Jemez Valley.[12]

CHAPTER 17

Penitentes in the Jemez Valley

Growing up in Northern New Mexico, I heard about the Penitentes, especially around Easter, when their secret rituals were sometimes performed. One of the longer and somewhat puzzling names for this religious brotherhood was Los Hermanos de la Fraternidad Piadosa de Nuestro Padre Jesús Nazareno (The Brothers of the Pious Fraternity of Our Father Jesus the Nazarene).[1] But I never knew there were Penitentes in the Jemez Valley long ago until I read a story about them in Richard Baxter Townshend's *The Tenderfoot in New Mexico*.[2]

Penitente history in New Mexico extends back at least to the early nineteenth century. After Mexico won independence from Spain in 1821, the number of priests serving in New Mexico decreased. Small communities would be visited by a priest perhaps once per year. In the absence of clergy, communities came together to memorialize the Passion of Christ through acts of penance, often including self-flagellation and mock (or actual) crucifixion of a selected member of the community. They gathered in small churches called moradas and marched through their villages carrying out acts of penance.[3]

Townshend's Penitente story from about 1876 took place in the "Mexican village of Jemez," a few miles north of Jemez Pueblo. It is clear that he is referring to what is now known as Cañon (sometimes called Cañoncito on old maps). Townshend, a British rancher turned trader, had arrived in the Jemez in 1874 with two partners and a couple of freight wagons filled with trading goods. They set up a trading post somewhere in Cañon. One of his partners, Vicente Elias, was murdered there one night over a dispute between "local political factions, which were very bitter" (see chapter 16). After that event, Townshend sold his part of the business and moved to Jemez Pueblo, where it was safer, to live with his friend John Miller, who was a government agent there.

In this story, Townsend recounts his experience watching a Penitente ceremony and his conversation about it with his friend Miterio Archuleta. Miterio and his family lived up the canyon in Jemez Hot Springs.[4]

Riding up to Jemez [Cañon] on one of these rare visits of mine I was a little surprised to see quite a crowd standing at one of the entrances to the village, who were clearly spectators keenly watching some affair of very special interest that was going on. So much I could tell from their still, silent, absorbed concentration on the thing, whatever it was.

And then, as I rode near, I was able to see over their heads from the viewpoint of my saddle something that fairly made me sit up. The open space in the middle of the houses was empty save for a procession of six figures; six men stripped to the waist and barefooted, each having the face and head muffled close in a white cotton wrap which hid his identity absolutely. The six walked, stooping forwards half double, with bent knees and long slow dragging steps, one behind the other; each figure held in its two hands a many-thonged scourge made of soap-weed, and at each step the figure raised the hands to the shoulder and brought the scourge sharply down its own back: all the six scourges were red, and the bare backs were red, and the white cotton waist-cloths below were growing red, too, with the blood. Erect and bare-headed with no cotton wrap to hide his face walked in front of the line a man in ordinary clothes. He carried an open book, and his lips mumbled some droning litany or other. Suddenly it flashed across my mind how that very day, just before I started, some Indian in the pueblo had said to me, "Español very mad to-day: do much fool thing to-day." Of course, this flagellation was what the Indian had meant. And now I remembered to have read of such things as Flagellant orders as existing away ever so far back in the Middle Ages. Obviously, these self-torturers were a survival, and I watched them with redoubled interest.

The spectators in front of me stood there in ordinary dress, nor had they their faces hidden, and the next moment I recognized among them Miterio Archuleta. He was a son of old Francisco Archuleta of the Jemez Hot Springs who with all his family were always our good friends; they had no sort of use for poor old Vicente's brutal murderers. I signed to Miterio, and he came up and put his hand on my saddle and we talked.

Chapter 17

"What's all this business?" I asked him in a low voice.

"You no sabe?" returned he with surprise. "These are the Penitentes." Of course, he too dropped his voice so as not to be overheard, for we spoke Spanish.

"But who and what are they?" I said. "And why do they do it? Do you know them yourself?"

"Oh, no," he returned, "though I can make a pretty good guess. They are a secret society, and they flog themselves thus every year. They meet in a big room they have in one of the houses here, and they know each other by secret signs and passwords; they call themselves Hermanos, Brothers, and it is quite true that they have their lodges all over New Mexico. They take it in turns to flog themselves, and when their turn comes there is no backing out; they have got to take it. But I believe it only comes to each brother once in a few years. Probably half the men that you see here now looking on in ordinary clothes belong to the Society and have taken their turn. I couldn't swear to it; I never was a Penitente, me, but I feel sure enough about them. This is how they are supposed to expiate their sins: perhaps Vicente's murderers are expiating theirs here and now."

Wonderful, incredible almost, is the mystery of human nature! That men should be capable of an atrocious crime and capable also of submitting to this extremity of self-torture as some sort of atonement for it! I felt utterly staggered.

"They won't be moved to confess their sins, I suppose," said I. "They won't give themselves away to the civil law?"

"But you don't understand," said the Mexican. "This has nothing to do with the law. This is part of religion. It is all quite different. It may be that they will go out of the pueblo to the Calvary there on the hill and there they may crucify one of them as a sort of sin-offering. But the law, never."

"Crucify him?" I exclaimed.

"Yes," he answered, "crucify him, with cords, that is, not with nails. You know at the Calvary there are three or four crosses made out of heavy wood poles cut from trees. They may tie him, the one chosen, or it may be the one who offers himself for sacrifice, to a cross and set it there upright with him on it for hours and do their flogging of themselves round it. But that's not done often and I don't know if they will

Figure 17.1. Charles Lummis's 1889 eyewitness and photographic account of a Penitente ceremony in San Mateo, New Mexico, included this image, titled "Crucifixion of the Penitentes."

try that to-day. As I told you, I'm not a Penitente, me, and I'm not in their secrets and no one outside the society can know about what they intend to do."

"But does the man, crucified die?" I asked.

"No" he returned, "not very often, that is. They take him down before it gets as bad as that, only there are times when they are too late and then there's nothing for it but to bury him."

"And there's no inquest, nor nothing?" I said. "It's all hushed up?"

"Why, of course," he said. "Didn't I say that all this is religion and quite outside the law. Why, the very man whose duty it would be to set the law in motion here is one of the biggest men among the Penitentes. That much I know myself. Do you fancy he would do anything against his brothers in religion?"

I felt the hopelessness of the position.

"If you wait a little," said Miterio, "there may be some women coming out. They do it too sometimes."

"Never!" I exclaimed horror-struck. "Look here. I must say good-bye. I'm going back to the pueblo. I want to talk to John Miller. I never dreamed of anything like this."

"Well, good-bye," said the friendly Mexican. "Come and see us at the Hot Springs when you can."

And with no more ado I turned my mare and rode straight back again to Jemez [Pueblo]. And there John Miller, who knew a thing or two about New Mexico, confirmed every word Miterio had said.

Townshend's shocked reaction and description reflect his Anglo, Victorian-era prejudices and sensibilities. Modern historians hold a better-informed and sympathetic understanding of Penitentes, recognizing that the brotherhood represented a unique expression of faith in remote, isolated villages, with practices of penance tracing back to old Spain. There has been much misinterpretation and misunderstanding of Penitentes since the late 1800s.[5]

Townshend's stories are remarkable snapshots in time of people and places in the Jemez Valley of long ago.

Figure 18.1. Image titled "Correr El Gallo—Sunday Sports in Southern California," from *Harper's Weekly*, March 3, 1877. (Wood engraving after a sketch by Paul Frenzeny)

CHAPTER 18

Correr El Gallo at Walatowa

A couple of years ago, we decided to raise chickens again, as we had done in a previous place we lived. We enjoy the fresh eggs and the peculiar pleasures of raising and caring for these quirky birds. So we acquired six chicks from a feed store in Bernalillo, and we built a coop and a fenced enclosure for them here at our place just north of Jemez Springs.

They were supposed to be all hens, but one turned out to be a rooster. He grew large and increasingly aggressive, but we tolerated him because he added a little charm to our barnyard. His morning crowing woke us up, and I suspect woke the neighbors too. The grandkids were amused by him, naming him Chile Peeper.

But he picked on the hens too much, treading out the feathers on their backs. It's a common behavior of overbreeding, when a rooster has too few hens and abuses the few he has. Aggression toward intruders is also typical. My wife had to carry a small rake to fend him off whenever she entered the coop to feed and water the chickens and to collect eggs. I went in one time to collect eggs and he attacked me from behind, drawing blood from my leg by stabbing me through my jeans with his two-inch spurs. Chile Peeper was a bad boy, but he thought that was his job.

The last straw was another attack by Chile Peeper on egg collectors. My son and family were here visiting, and my son took charge. Among his various talents, he's a very experienced wild game hunter and a gourmet cook. He quickly dispatched Chile Peeper and then scalded, plucked, gutted, quartered, and slow-cooked him in a crockpot. More about that outcome later.

As this ancient ritual of harvesting livestock was going on, I recalled another use of tough roosters that is also relatively old in the Southwest and the Jemez. It is known in the Spanish-speaking world by various names, and it has different traditions by place and time. But here in New Mexico, among old Hispano and Pueblo cultures, it is called Correr El Gallo, or Running the Rooster. It was a common event on San Juan feast day on June 24.[1]

Now, I feel obliged to issue a trigger warning to those who would rather not read about a game that many consider cruel to El Gallo. If that's you, I suggest you stop here. Still, I will note that uncomfortable history sometimes provides insights about people, places, and times past.

The typical Correr El Gallo in the greater Southwest, including southern California, Arizona, New Mexico, and northern Mexico, involved burying a tough old rooster in the ground, with only its neck and head sticking out. Then young men on horseback galloped by while trying to grab the bobbing and weaving head of the live rooster. The rest of the game played out in various ways but usually as a mock battle and tug-of-war. In sum, this was a very public contest of horsemanship, toughness of the young combatants, and the resilience of the poor rooster's corpse![2]

The following excerpt is a description of a Correr El Gallo event at Jemez Pueblo (Walatowa) in the summer of 1903 as witnessed by the British classics scholar and one-time resident of the Jemez Valley R. B. Townshend:[3]

> But before I had got my camera loaded and cocked there was a cry from the Indians, a yell I might call it, though not so blood-curdling as the war whoop. A white shirted rider had grasped the object in the sand as he stopped, and out from the sand in which it had been buried came the form of an elderly cock. The rider recovered himself into the saddle with a shout of triumph and waved his trophy aloft—I think its neck was already broken by the jerk—and he dashed away for the edge of the Pueblo. The others [went] streaming after him at full gallop, the horses manes and tails flying and the gay shirts and the leggings of the riders fluttering in the wind. I ran to the top of the little hill by the corrals, but I was too late to get a [photo]; they were already outside the village, rushing along the rough ground at the foot of the low mesas which border it. [They were] strung out like a Rugby football team after the man who has got the ball. For that is how they play the game, only that it is on horseback, and the coveted prize carried by the rider is the body of "el gallo" and not an ellipsoid of brown leather.
>
> But if I was too late at this time to snapshot the start of the game, I found no lack of subjects. For though there were some sixty

Chapter 18

Indians, perhaps, on horseback, they did not all play at once. Not more than a dozen or twenty had gone in pursuit of the bearer of "el gallo," and the others reined in their fiery ponies and [only] looked on for this bout.

What splendidly picturesque groups they made there on the knoll in the bright sunshine. They were all young, the young braves of the tribe, from seventeen to thirty years old perhaps. Each had his long hair tied up in a thick club or knot bound with braid hanging at the nape of his neck, while his black sidelocks were kept from blowing into his eyes by a red fillet or twisted handkerchief bound tiara-wise round his brows.

Their gaily colored shirts were gathered like blouses round their waists by a cartridge belt and the skirts fluttered free below, for the Indian wears his shirt outside his leggings. Few of these had the true, old-fashioned red buckskins of their ancestors, but many wore instead the trousers bought at the store of the American, and in order to give them a proper Indian effect they had taken in a broad tuck all along the outside seam, making them cling close to the leg in proper Indian style.

Others wore the short wide loose trousers of white cotton cloth coming halfway down between the knee and ankle, which are quite as typically Pueblo Indian as the buckskin leggings, and are certainly no whit less picturesque. On their feet were red moccasins of their own special make, the "teguas" of the tribe. . . .

I stayed right with the crowd of "gallo" players to see what they would do next. . . . I could not translate exactly what they said, but I think they talked of the runners in the scurry which had just finished, and who had been well up, and who had been tailed off in the race. What I think happened next was that Blue Shirt said to White Shirt "Yes, you came in first, thanks to your father's good pony, but if I'd been able to get up close enough to collar, I don't think you'd have kept that gallo." And then I fancied the White Shirt retorted, "If you think that, you'd better come and take it now while we're standing still." And the backers of White Shirt cheered him, while Blue Shirt's friends bade their champion to wade in.

Anyway, Blue Shirt settled himself in his saddle and grasped the long ends of his reins in his two hands, which he held about a foot apart, straining the leather thongs lightly between them, while White Shirt, holding the gallo firmly by the legs, swung the white body of the rooster up over his shoulder and made ready for the fray.

With his heels Blue Shirt urged his pony up to White Shirt's, holding up the strained leather as a guard to his face. As they closed the blow descended and White Shirt banged Blue Shirt over the head with the body of the gallo with all his might. Blue Shirt bored-in to grapple, his hands up guarding his bent-down face, and taking the rain of blows on his head and back. A good heavy rooster, wielded by a vigorous arm, can give a smashing blow. The feathers flew in clouds, and flecks of blood (the defunct rooster's) spotted the clothing and the pony of Blue Shirt.

The sight of blood in a physical contest stirs the natural man and the shouts of the backers were redoubled. They urged their ponies right up against those of Blue Shirt and White Shirt, driving the pair to closer quarters. And now Blue Shirt had his chance, for thrust thus upon his adversary he succeeded at last in grasping the neck of the gallo as it descended. Instantly the shower of blows stopped, and the contest became a tug-of-war.

Each rider pulled against the other with all his might, and the press of the other riders opened out to give them room. The principal backer of each slashed his man's pony over the quarters to make him break away. Sharp cries and yells rent the air as Blue Shirt and White Shirt threw themselves right over the sides of their ponies to get a full purchase against each other.

Would the gallo come in two? If he had been a young bird he must, but the respected parent who has lived to see several generations of his tender spring chickens come into being and pass through their destined course of a short but happy life to a swift and savory end, is himself of tougher quality. Age which unfitted him for the pot gave his sinews the resisting power which made him suitable for this game. Pulling, hauling, go White Shirt and Blue Shirt, but the gallo still holds out, though perceptibly growing longer. "I may stretch," his spirit might say, "but I won't break."

But something has to give, for the ponies are being urged to part with vigour. Blue Shirt with only his right leg visible over the side of his pony next his adversary has got the gallo's neck against the horn of his saddle and holds it there with a desperate grip. The pony's pull drags White Shirt, who has also been hanging over his horses side, [gets] up onto his saddle again. In the next moment White Shirt is over on the near side of his pony and has to let go to save himself. Blue Shirt wins and his triumph is heralded with a staccato chorus of "Hi-hi-hi-ah's."

It was a famous victory, and the winner holds himself as proudly as if he were Wellington after Waterloo. And his side—for it seems as if there were sides, though I could not guess on what principle they were organized—laughed and shouted an Indian version of "Wha daur meddle wi' me?" [an old Scottish phrase, loosely translated as "No one can harm me unpunished"].

Well somebody does "daur meddle," it appears for White Shirt was not satisfied with his defeat, but made him ready for the attempt to win the gallo back. Up he came to the scratch gallantly, guarding his head with his reigns as Blue Shirt had done. Blue Shirt whirled up the rooster, longer now and more limber for all his stretching, and poured it into White Shirt over his guard.

The rain of blows was such that the red fillet round White Shirt's head fell off, and even the strong braid that bound his locks behind into a club came undone, so that his long, glossy black hair fell streaming down to his waist and made him look wilder than ever. But he took his punishment stoutly, and bored-in on his opponent, till at last he got a grip on the gallo's body and checked the storm of blows, and thrusting one hand forward seized one of the legs which were tied together with the cord.

Down behind their saddles the two youths flung themselves, their hands overlapping as they pulled with desperate grip. For now, they had each got a leg of the gallo to hang on to, and it was a question whether the tug-of-war would not end in a dissolution of partnership. And that is exactly what happened, for with a violent wrench White Shirt tore away the leg he was holding, the cord broke, the leg came away, and the parting of the sinews was so sudden that he lost his knee-grip over his pony's back and plopped headlong to the ground. He was up in a moment and brandished the trophy he had secured as Blue Shirt swung up the larger portion to renew the fight.

But it was settled by the others that it was poor sport to go on with a dismembered gallo. The pieces were passed on to the younger boys to practice on, and the band of horsemen rode down to a certain house in the village, and called out loudly, "Otro gallo, otro gallo!"[4]

I doubt that two-year-old Chile Peeper would have stretched and lasted as long as the older rooster in the 1903 contest. He was, however, an impressive plucked bird with very muscular and fatty legs and torso. I thought he would be rubber-tough chewing, but the meat was savory and quite tender after slow cooking for nine hours using a French recipe for old chickens—the coq au vin style—a red wine, mushroom, vegetable, and spice stew. The taste was nothing like your usual chicken. It was rich and flavorful, like a combination of dark turkey meat and pork roast. Still, if we ever harvest and cook an old chicken again, with a nod to my son's culinary skills, I think we will try a slow-simmered adobo stew using Jemez red chiles.

CHAPTER 19

Presbyterians in the Jemez

The oldest church in Jemez Springs is the Mission de San José at Gíusewa, now in ruins, built between 1621 and 1626 (Jemez Historic Site). This Catholic church was probably abandoned by the Franciscan friars before 1650 due to Navajo, Ute, and Apache raiding.[1] The next oldest church in Jemez Springs is the Presbyterian church, dedicated in 1881. How did a Protestant church establish here in this remote village of longtime Spanish and Catholic origins? There are multiple historical reasons.

Following the Civil War, Presbyterian missionaries established churches and mission schools in towns and small villages in New Mexico. For example, the founders of the Menaul School in Albuquerque were leaders in the Presbyterian mission school movement in New Mexico. Many New Mexicans had fought for the Union during the war, and by the 1870s some of their children had been educated in Presbyterian mission schools in New Mexico or the East, where they were exposed to Protestantism. Their education included access to both Spanish- and English-language versions of the Bible. These changes prompted some families to leave the Catholic Church and join the Presbyterians, at least temporarily.[2]

Another reason for the spread of Presbyterianism in New Mexico was a Catholic schism that developed after the Americans arrived in 1846. Conflicts arose between New Mexico–born Catholic priests and recently appointed priests from France. In 1851 the new bishop, Jean-Baptiste Lamy (later archbishop), began trying to suppress the Penitentes. He also imposed mandatory tithing and other unpopular policies. Willa Cather portrays aspects of the Catholic turmoil in New Mexico in her famous novel *Death Comes for the Archbishop*.[3]

In Cather's fictionalized account, Lamy and the French priests he appointed are portrayed favorably, but a character named Padre Martinez of Taos is thoroughly corrupt. In reality, the French archbishop and his appointees were often condescending and offensive in their attitudes toward old Spanish

Figure 19.1. John Milton Shields and family in the early 1880s. (Courtesy Adams family)

Chapter 19

New Mexican culture. And the actual Father Antonio José Martínez of Taos was a visionary reformer.[4]

Padre Martínez was indeed a controversial character in New Mexico politics for decades. Still, he was also a champion of the Penitente Brotherhood, an advocate of coeducation, and progressive in other ways that defied Lamy's orders and policies. For his troubles, he was excommunicated by Lamy. After Martínez died, one of his sons (whether biological from before he was ordained or adopted is not known) joined the Presbyterians in Taos and was made an elder in 1872.[5]

Several Presbyterian missionaries and families found their way to the Jemez Valley in the 1870s to 1890s. The Reverend John Milton Shields, his wife, Emily, and their two sons traveled from Pennsylvania to Jemez Pueblo in the spring of 1878. Shields was a Civil War veteran and a medical doctor. The people of Jemez Pueblo had suffered greatly from a smallpox epidemic during the 1870s, so a physician was welcomed.

Emily died of illness at Jemez in November 1878. Shields then married another missionary from Pennsylvania, Isabella Leech, who had arrived at Jemez in 1879. They had eight children.[6]

In 1881 Francisco Perea moved with his large family to Jemez Springs, where he managed the newly built Stone Hotel; that building is now part of the Bodhi Manda compound. He was also the village postmaster from 1894 to 1905. Perea's connection to the Presbyterians was personal and familial: He and his immediate family were Presbyterians, and his brother José Ynez Perea was the first ordained Presbyterian minister of Spanish descent in New Mexico. José Ynez was involved in establishing the Jemez Pueblo mission with Shields in 1878 and the building and dedication of the church and mission school in Jemez Springs in 1881.[7]

Francisco Perea, you might recall from previous chapters, was the son of a wealthy and politically influential family of Bernalillo. He had been a Union Army colonel in the Civil War, a territorial representative in Washington, DC, and a friend of Abraham Lincoln. The Stone Hotel and the bathhouse in Jemez Springs were built and owned by Mariano S. Otero, a wealthy businessman, sheep rancher, and politician. Francisco's uncle José Leandro Perea Sr. owned a mortgage on the Baca Location in 1875, and by 1899 Mariano Otero had purchased the entire Baca. (These lands are now the Valles Caldera National Preserve.)[8]

Figure 19.2. John, Hugh, and Mary Stright Miller at their home across the road from the Mission de San José ruins in Jemez Springs, 1903. Photo by R. B. Townshend. (Courtesy Pitt Rivers Museum, University of Oxford, Catalog No. 1998.58.55)

Chapter 19

Both business and marriage connected the Otero and Perea families: Francisco's first wife, Dolores, was Mariano's sister, and Mariano had married a Perea cousin. The Oteros also lived for a while in Jemez Springs after 1882. Their house was probably the original part of the building now owned by Jemez Pueblo, which was occupied after 1947 by an order of Catholic nuns, the Handmaids of the Precious Blood.

Another Presbyterian missionary from Pennsylvania, Mary Stright, arrived in 1880 to assist in the mission at Jemez Pueblo and the new church and school in Jemez Springs. She married the longtime Jemez Pueblo government agent John Miller, and they moved to Jemez Springs, where they built a large home across the road from the old Catholic church ruins at Gíusewa. Shields and his family also moved to Jemez Springs and were next-door neighbors to the Millers. One of the Shields daughters, Helen, married John Amos Adams, who was a surveyor for the US Forest Service in the early 1900s.

A longtime Presbyterian family with descendants still living in the Jemez Mountains are the Fentons. The Reverend Elijah Maclean Fenton Sr. arrived in 1894 from Missouri. He and his wife, Jessi, owned land in Jemez Springs, but their main homestead was on the Rio Cebolla, near what is now Fenton Lake. They built a sturdy stone-walled house and raised a family there (see chapter 20). The Reverend Fenton presided at the Presbyterian church in Jemez Springs on many occasions. Their son E. M. (Mac) Fenton Jr. was well-known in the valley in later years, as was Mac's daughter Mary Fenton Caldwell and other descendants.

Also living on the Rio Cebolla were Juan and Juliana Sandoval and family. Juan and Juliana were baptized as Presbyterians in 1880, with Shields presiding. José Ynez Perea was also present at the ceremony. Juliana's father and mother, José Francisco Archuleta and Maria Viviana Montoya Archuleta, were confirmed Presbyterians living in Jemez Springs. They donated the property where the Presbyterian church and cemetery are located. Their son and Juliana's brother Emiterio (Miterio) Archuleta was an elder in the church. He was also a neighbor of the Millers and the Shields on the lands south of Soda Dam extending to the old mission church ruins.[9]

The most colorful descriptions of people and places in the valley during the 1870s and 1903 come from Richard Baxter Townshend's *Tenderfoot* books. In 1903 Townshend visited the Millers, Pereas, and Archuletas in Jemez Springs. And at their invitation, he visited the Fentons on the Rio Cebolla. He also visited the Sandovals on that trip and took photographs (see chapter 20).

Figure 19.3A–B. Francisco Perea as a young territorial representative in Washington, DC, in the 1860s and late in his life. He lived and worked in Jemez Springs as a hotelier and postmaster from about 1881 to 1910. He died in Albuquerque in 1913. (First photo courtesy US House of Representatives; second photo from Twitchell, *Old Santa Fe*)

The following are excerpts from Townshend's 1903 "Letters from Jemez Hot Springs,"[10] written to his wife, Dorothea, back in England. Townshend's Victorian-era, British-Anglo prejudices and predilections are evident in parts of his letters. He describes his impressions of the Presbyterians of the Jemez Hot Springs congregation. He expresses surprise at how much things have changed in the valley since he was here as an adventurous young man in the Wild West days of the 1870s. These feelings are familiar to those of us who lived here long ago in our youth, went away, and then returned—a mixture of nostalgia for the old and wonder at the new. As the novelist Thomas Wolfe said, "You can't go home again."

When I had written my last to you about the old church and the San Diego Cañon, I went down with John to post it, and found that the post-master was at prayer-meeting! It seems they have a Presbyterian church here, and they use it, too, and several Mexicans have joined. Well, we turned into a store, and John smoked, and we chatted for an hour or so with the storekeeper and various Mexicans, and then sallied forth as prayers were over.

And when I got into the post office who was the postmaster but Colonel Francisco Perea! The man who said, "Wait!," the cousin [nephew?] who was opposed to Jose Leandro. He is poor! and lives here humbly with his family. He is old and dim-sighted, and it was only with some difficulty that he found your letter for me. But he knew me at once, and referred to old times, and I must go down and have a chat with him. It seems the Pereas have all lost their money somehow and gone scattered.—Wonderful. I'll find out more and tell you if I can get anything interesting. I suppose it is partly the influx of American competition.

Next day we all walked to the Presbyterian church in the blazing sun. Oh, I don't know how to express things at all. Here where I was a member of that somewhat semi-civilized community we have two churches, one Roman Catholic, one Presbyterian! And who are the Presbyterian elders? Colonel Francisco Perea (who said "Wait!"), and Miterio Archuleta who helped me drive those steers to Santa Fe, and as a gay young spark then. I sat in a chair in a little whitewashed barn sort of place, and Miterio came and gave me a hymn book in Spanish, and Mrs. Miller played the wheezy harmonium, and we made a joyful noise.

Well, I need not say much more. It used to be the Dark Ages. Flagellants, witches, and murderers! And now the hymn and old Dr. [Shields] preaching. It would take a Hardy to give the curious irony of it. I guess it had to be. I guess it's for the best. Yet if I could express the change in words, I'd make folks feel something!

I sat in chapel alongside Mrs. Miterio, old now, one-toothed, grizzled, parchment face. I remember her young, charming; would she say these days are better than those? Her face was non-committal; but she came all right to church, and her children and grandchildren. Valgame Dios! If I don't feel as if I was standing on my head!

And John sat in front of me, shaved, and in clean shirt and a black coat, and the young Mexican population around looked bored or tolerant, or whichever it is, including two stalwart sons of Juan Sandoval. I tell you I thought things. For we didn't go to church much here thirty years ago—you'd see the young fellows put for the hills when the padre came; and now I wonder to myself, are the old men and the young men two nations? For it feels as if they were. Or does it mean that there is a real change and this little Dr. [Shields], grizzle-bearded, mild, spectacled sixty, who came out as a medical missionary and teacher to Jemez in 1878, talking bad Spanish through his nose about "La Palabra de Dios," represents a winning force? For all things here seem changed.

The men who were the well-to-do men of the Cañon de Jemez when I was here are mostly poor now. They had sheep on shares from the "ricos" of the Rio Grande and seemed to be doing well. Then John tells me there came a year when the pasture was bad in the low country where they used to winter their flocks while there was still good grazing in the mountains. And they moved up their flocks earlier than usual, about the beginning of April. Perhaps 100,000 sheep went in. And then

Figure 19.4. (*opposite above*) View looking north in Jemez Springs in 1884. The numbers identify owners and buildings: (1) Home of James Smith and Calletana Archuleta Smith (previously the home of José Francisco Archuleta and Maria Viviana Montoya Archuleta); (2) Unknown home; (3) Otero bathhouse; (4) Otero Stone Hotel; (5) Unknown home; (6) Home of Francisco Perea and his second wife, Gabriela Montoya Perea (now the location of the Los Ojos Bar); (7) Jemez Springs Presbyterian Church. In the far distance, the bell tower of the Mission de San José at Gíusewa (Jemez Historic Site) is barely visible. Identification of the owners and buildings are based on Jemez Springs Presbyterian Church records, newspaper articles, old photos, and the 1992 Our Lady of Assumption annual fiesta booklet. (Courtesy Library of Congress, LC-USZ62-113676)

Figure 19.5. (*opposite below*) Photo taken with a drone with approximately the same view as the photo in Figure 19.4 but elevated above the trees to show the old and modern buildings. The Jemez Springs Presbyterian Church is visible at the lower right (with the newer roof peak, oriented east–west versus north–south in the 1884 photo). The bathhouse and library, at left center, have gray roofs. The old Stone Hotel, now a Bohdi Manda (Zen Buddhist) building, just below the line of trees at upper center, has a white roof.

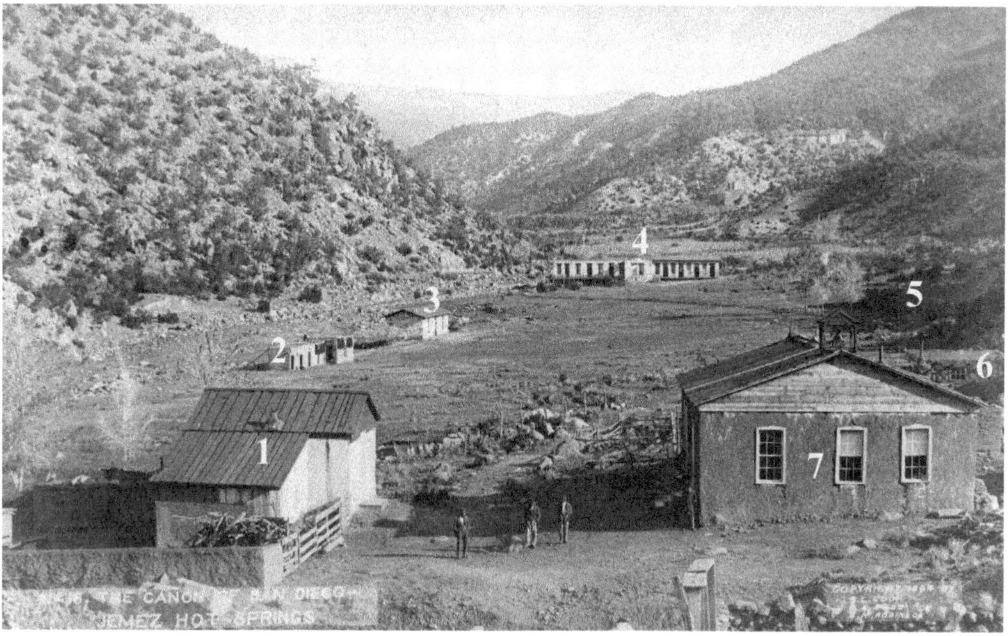
4
3
5
2
6
1
7
THE CANON SAN DIEGO
JEMEZ HOT SPRINGS

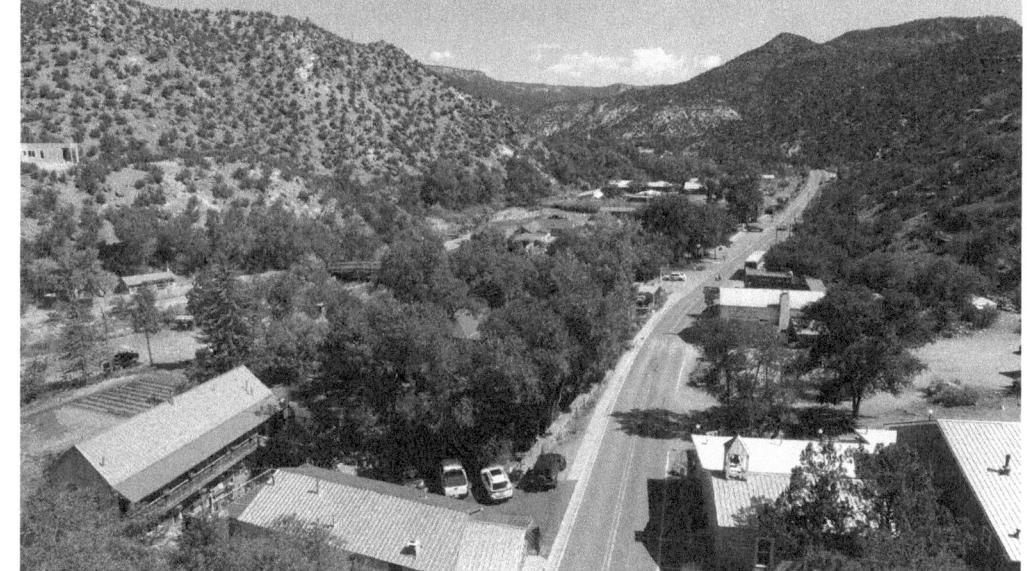

the snow came down for three days and nights and lay three or four feet deep on the level and the sheep were smothered wholesale.

Men who went in with 4,000 sheep came out with 1,400, and lucky at that. Those that survived barely paid back the original stocks to the "ricos," the great flock-masters, and these farmers who had most of them debts to pay and were liable for the balance if their herd did not fill the original bill, had to sell their little bits of land to clear themselves. So, it comes that when I ask after this man or the other, Andres Archuleta, Christoval Cassados, Clemente Cassados and even Juan Trujillo, who seemed himself to be even as one of the "ricos," I hear, "Oh busted," "Living very poor in Nacimiento [Cuba]," "Living in Albuquerque," and the like . . .

But as for me, who thought this "rincon," this most out-of-the-way part of the world, would remain as it was? I can only say it makes me feel topsy-turvy. And that Presbyterian service this morning has put the finishing touch on it. . . .

I like to see the old mountains again, though they are changed, for the washing of the rains has altered the valleys so that I would hardly have known them again. But still the air is there and the forest. But for me who remembers other things it's all different. For instance, poor old [José Francisco] Archuleta! How things have changed since I knew him! [I] Think of him in his lusty youth, running over these mountains in moccasins, bow and quiver at his back, slaying the wild deer and even mountain lions with his arrows . . . he killed four lions one day with arrows! Aye, and Navajos, too!

And then in a mountain he found his two brave boys, waylaid, and killed by the Navajos in turn, and wept so that he went blind, and his horse brought him home. And then in his old age came the new days, and he saw the railway, of which he had only heard by report, and the coming of the Americans.

Yes, I'd like to have seen him again, but I dare say he is better sleeping in quiet earth. But he was a grand old hunter, warrior, scout of the mountains, a true-bred son of the Conquistadores. Peace to his ashes.

CHAPTER 20

A Visit to the Rio Cebolla in 1903

Richard Baxter Townshend was a keen observer, and he wrote with strong feeling about this landscape and its people. The Jemez clearly made a deep impression on him. His books are full of stories and heartfelt letters home to his wife, Dorothea, in England. He tells stories about his time living at Jemez Pueblo, hunting, and hiking on the mesas. His books also include stories of ranching and mining adventures in southern Colorado, rabbit and horse drives, an encounter with Billy the Kid, and witnessing a snake dance at the Hopi Mesas.

Townshend was a man of his time, growing up in Victorian-era England, and his character was shaped by his youthful adventures in the Wild West. The excerpts from his books that I included share his observations and reveal his prejudices. Fair warning if you read his books: his prejudices sometimes stray over the line into racism.

Despite this, and to his credit, Townshend had considerable empathy and a perceptive recognition of the mistreatment of people. For example, on his visit to the Hopi in 1903, he described and condemned US government's forced removal of Indian children from their families for "education" at far-off boarding schools.[1]

The following is another excerpt from Townshend's "Letters from Jemez Hot Springs," written to Dorothea during his last trip to the Jemez. This story derives from Townshend's visit to the homes of the Reverend Elijah Maclean Fenton and his wife, Jessi, and Juan Sandoval and his family, all living on the Rio Cebolla. The "broad green meadow" by the Fenton home is now Fenton Lake. He traveled there by horse-drawn wagon with his old friend John Miller and Miller's young son, Hugh. The route they followed north from Jemez Hot Springs was mostly along what are today NM 4 and NM 126.[2]

July 15, 1903

John and Hugh and I set off to the mountain in his buggy. We went up to the Valle de la Cueva, the same road as I went with Mr. [Block], only it seemed more natural in a way to go with John, though there is a difference between our trip of twenty-seven years ago and to-day up the same cañon. I walked from the sawmill to the Cueva so as to get a good chance to see the views up the gorge, which are really very fine. I am sure lots of people would come to see this country if they only knew about it.

At the Cueva we sat by a little spring and ate a lunch of cold flapjacks and butter and cookies and pie, and then we cut across the creek and up some astonishing pitches to get back to the road to the Fenton's. Do you know, I had quite forgotten what horses and wagons are capable of. Really the way they climb is incredible. I quite understand how our fellows in S. Africa pooh-poohed the idea of Boers being able to get cannon up on to lots of kopjes and then found themselves peppered from these same inaccessible spots. Boers drive like American Western men.

Once on the road we wound up and up spurs of the high mesa till we must have been one thousand feet above the Cueva. I looked back through vistas of splendid red-stemmed pines and dark-stemmed spruce and saw away below the green bottom of the Cueva—green now, alas! with weeds only—the innumerable prairie dogs have ruined the grass—and beyond it the great mass of the San Antonio Mountain rising grandly in great rounded domes—it is not craggy—dark with pine and light green with mountain aspen and wild oats, with purplish patches of screes here and there to the very summit.

And then we reached the mesa summit and found ourselves in even finer pine timber and with a much more lush growth of underwood. For some reason this side of the mesa was damper and the trees more dense, and presently we entered the upper end of a long, gently sloping cañon which led down to the Cebolla, and here in the cañon were Juan Sandoval's two sons lying beside their two stray horses close by.

We said "Buenos tardes" and trotted on, but presently they mounted and came galloping past us through the bush, driving before

Figure 20.1. The Fenton home on the Rio Cebolla. The girl is Jean Fenton, the boy is Hugh Miller, and the dog is Tim. John Miller and Jessi Fenton can be seen sitting on the porch in the background. Photo by R. B. Townshend. (Courtesy Pitt Rivers Museum, University of Oxford, Catalog No. 1998.58.91)

them colts they had been to fetch. And our horses began to gallop, too, though with discretion, and finally they went ahead of us and disappeared in a cloud of dust down the cañon. The cañon grew deeper as we went on, and the cream tufa cliffs on our right were wonderfully carved and cleft into tall cave-mouths, some of which ran far back inside. Juan Sandoval lived in one for a time, said John, before he fixed up his ranch.³

    And then we came out into an open space where was a cabin and a corral and cows and two young men and an old one, and the old one came up and shook hands, and it was Juan himself. The women-folk came out of the cabin to look at us, but we were late and didn't stop, but waved our greeting to them and went on over a little low divide to the next cañon quite close, and here was Mr. Fenton's substantial stone house and a broad green meadow with a thick, broad line of willows winding through it and this was the Cebolla.

We drove up to the house, and cheerful little Mrs. Fenton greeted us, and we got off and unloaded and turned the horses loose and sat on the broad stoop and talked. It was sunset and the air had the keen fresh mountain sharpness in it, so startling a change from the lower levels. A thousand feet between Albuquerque and Pueblo de Jemez doesn't seem to make much difference, while six or seven hundred feet between the pueblo and the Hot Springs give one a new climate and coolish nights, but the last thousand feet between the Hot Springs and the Cebolla give one the true breath of the mountains.

I wish you could breathe this; I know you would like it. Alas! one mile of these roads would kill you. By the way, I stood it all right yesterday though I am sore and bruised from sitting with the lower part of my spine against John's Winchester. Also, I walked about two hours out of six. But I had four hours of jolting.

Well, after a little while on the stoop, during which I made friends with Tim, a big, tawny, half-bloodhound, half-collie dog with a splendid deep voice—he favours the bloodhound—Mr. Fenton hove in sight. The Fentons have two children, Jean, the dear little freckle-faced girl that came down to the Hot Springs and rides like a centaur and does heaps of work round the ranch, and Maclean, a boy of about Hugh's size.

We had a sumptuous supper of eggs and flapjack and lettuce and mush and maple syrup and gooseberries, with napkins, and I had a good bed with sheets, and Mrs. Fenton declares apologetically that they are living like rude pioneers. "We didn't have napkins and sheets," I tell her, "thirty years ago."

I forgot to say that it seemed kind of natural to see Fenton come riding in with a gun, a Winchester slung to his saddle. I have seen about three guns slung since I hit the Picket wire [Purgatory River in Colorado], instead of every other man wearing one as a matter of course [in the old days]. We inquired what he carried it for—John has his along,

Chapter 20

too, as my backbone can testify—John says he won't allow "no buck to bite him"—a staple "ranch" joke, now that a ranchman is restricted to one legal deer per annum.

Mr. Fenton carried his for prairie dog. He got him, too. You see, they have ruined the Valles on the San Diego Fork of the Jemez River, but they are not on the Cebolla yet, and if he can keep them out, he will. One colony started, but he and John shot them out. Yesterday, however, he suspected a prairie dog pioneer of spying out the land, and sure enough he got him.

I guess Mr. Fenton is quite Cromwellian in his ability to shoot as well as pray. You know it was quite novel, and I liked it, this morning when after breakfast we shoved our chairs back and he read a chapter (read it well) and prayed. I think he's an awful good sort, and he seems an uncommon good man with horses....

But as we talked of prairie dogs and the importance of killing them out (they are like the Australian rabbit pest)... Mrs. Fenton told in her lively way how when John Miller and Frank, her brother, killed all those prairie dogs up above here, they declared they were good to eat, and they dressed them and insisted that they should be cooked. And Mrs. Fenton and her sister demurred and didn't want that stuff in their frying pans. But the men insisted on it.

So, the ladies went to work and cooked up a quantity of prairie dog in the most delicious and titbit style, and they were to have it at (midday) dinner. But after breakfast Frank said he believed he'd go fishing, and as he might be lucky and not want to leave off, he'd just take a cold lunch along. And John Miller said it was fishing day for him too, and he liked to feel independent about meals when he fished. And Mr. Fenton said he thought he'd go off and see Rogers about cutting some hay. (Rogers lived ten miles off, and it was six weeks before hay harvest.)

And then Mrs. Fenton said, "No, you don't. You don't take any cold lunch with you, either of you, and you don't go off to see Mr. Rogers. You stop right here to dinner, and you eat dog!" And they did, and they swore they'd never dreamed of evading it, and that it was perfectly delicious, and they quite liked it!! ...

The Jemez Mountains

Today it was decided to have a picnic up the river, starting about ten, and then Mrs. Fenton and the children were to ride on some five miles to call on some new settlers, named Rouse, while we fished. So, John set out to fish at once—he's keen as ever on it—and I took my camera and sallied out up the bench alongside Barley Cañon, a side cañon that runs just above here. I do love a ramble with my camera in these mountains. The possibility of getting a shot gives it just enough object to make one feel happy....

I don't know the Highlands, but [Sir Walter] Scott seems to me to be always describing country like this. These cañons here are just glens. They are fine wooded valleys with the sides, say, one-third bold crag and two-thirds shaggy forest. They may be five hundred to one thousand feet deep and a hundred yards to a quarter of a mile wide. They wind and fork and have side cañons coming in, but the striking thing is the abundance of timber and brush and grass and wildflowers. I do like walking a few steps and sitting down on a fallen log and enjoying birds and butterflies and then rambling on little further and sitting again. Every time you get a fresh view and fresh beauties and drink the mountain air.

You will understand the way I feel about it, though I can hardly put it into words. And this part is unspoiled still, though there are fences here and there to restraint the stock, and in Barley Cañon a patch of arable [land]. Do you know, I don't resent the alteration of this from the days of the wild men, or even of the nomad Mexican flocks. I feel it had to come, and it is no more to be sighed over than over one's lost youth. Change has to be.

What I am thankful for is that I saw it yet unspoiled and felt the charm, and I tell you it makes an immense difference to me that coming back to it now I am with nice people like the Millers and the Fentons and Miterio and the Sandovals. Well, if I'd found that Americans of the ruffian type had come in, I'd simply have had to flee, I couldn't have stood it. But I can't tell you how I like a few hours stroll in the wild wood. I lack nothing but you.

Then Townshend describes the Fentons in admiring detail:

Chapter 20

I incline to think that this sort of Presbyterian preacher is sport enough for anything. He's from old "Missourah," six feet if he's an inch. When he courted the present Mrs. Fenton she was at the then Presbyterian Mission at Los Corrales—you remember—by the Rio Grande where the driver Tomas Griego, drove me through the flood-water after leaving Albuquerque.

Well, there was no bridge in those days, only the old bridge nine miles down, and Mr. Fenton, meaning to get married arrived on horseback at Alameda to find the river up (with ice-water too, in April, if you please) and the man who had the boat—they had a boat for high water then—disinclined to cross at night. The long Missourian remarked that he intended to get married, and left his horse with the ferryman, and swam it in his clothes, and reached Corrales on time, but so paralysed with the cold that he could scarcely walk; whereupon the Rev. Mr. (name forgotten) produced brandy and compelled him to take a quantity of it—the first liquor Mr. Fenton had ever taken in his life—but failed to make him intoxicated!!

Imagine if you can the verve with which lively Mrs. Fenton told me all this. They got married all right, and the horse arrived next day, but I guess Jean has a pretty good hereditary right to her courage. Fancy that chit of a child running the ranch when her father is away and separating those squealing, striking horses.

Did I tell you Kit and Prince had a fight? Prince was a stallion till he was four years old, and has heaps of fight in him. She separated them, tied up Kit in a stable, and put Prince in the burro pasture. Mrs. Fenton tells me Jean says to her, "I suppose I ought to help you, ma, with the dishes, but I'd rather go out and look after stock."

One time Mr. Fenton let out that he wanted the whole heard of cattle corralled including a fighting bull, which he has since disposed of, who would go for anybody if driven. Jean presently slipped out, got her horse, rode around the cañons, corralled the lot, and then came back to be scolded, knowing full well she would not have been let go if her purpose had been suspected.

Her aunts in Pennsylvania (where Mrs. Fenton comes from—the Millers, Shields, and Mrs. Fenton are all Pennsylvanian) want her

Figure 20.2. *Left to right*: Maclean Fenton Jr., Hugh Miller, and John Miller at the Fenton home on the Rio Cebolla. Photo by R. B. Townshend, 1903. (Courtesy Pitt Rivers Museum, University of Oxford, Catalog No. 1998.58.90)

sent back there to be properly educated, but she hates the idea, and, of course, Mrs. Fenton feels it would be a great separation. I encouraged her in the idea that Jean would educate herself at home—she's a great reader—and that home counted for more than schooling."

An epilogue: It seems Jean did not go off to school, but if she did, she was back on the Cebolla by 1915, when she wed Albert Phillip van Derveer. They lived there and had two children, a girl and boy. Sadly, the infant boy died in October 1918, and then Albert died in January 1919 of the Spanish influenza. An Albuquerque newspaper story reporting Albert's death noted that after a Christmas gathering of neighbors on the Cebolla, "several guests were stricken with influenza."[4] Jean ultimately lived a long life, and in 1981 she died at age eighty-nine in California, where she lived with her daughter.[5]

CHAPTER 21

Horse Logging Above Jemez Springs

There is a long history of logging in the Jemez Mountains. Some of the earliest logging was done with horses in the 1880s on the benches above Jemez Springs and below the cliffs of Virgin Mesa. Benches are the relatively flat areas below the cliffs in multiple places, such as along Ponderosa Drive on the west side of San Diego Canyon, known locally as Area 3.

One day in 1903, during Richard Baxter Townshend's visit to Jemez Springs, he took a hike north of Soda Dam up onto the bench where we live now in Area 3. He had been here in the 1870s, when the forest of big ponderosa pines was uncut, a so-called virgin stand. He wandered among the tent rocks and pine trees, taking photographs with his Kodak. This is how he described what he remembered from decades earlier and what he saw that day. His writing reflects an eye for the aesthetic and a poetic style you might expect from a classics scholar:[1]

> Here were the "monuments" of which John had spoken, an acre of rock pyramids, all sticking up close together, like the spines on the back of one of those wonderful horned toads, innumerable, sharp, many-fingered peaklets, the visible image of the "Mountains of the Globe" as figured on the first page of old-fashioned atlases. And there were others like haystacks, and one like a cardinal's hat, strange forms of rock carved by wind and weather, when strata of different hardness come together.
>
> I made sun-pictures with my camera and wandered far along the bench seeking more of these natural sculptures. One is always expecting to find a freak that shall be a perfect work of art. But works of art do not grow by nature. They are made, and man is "poietes," the maker.

Figure 21.1. Dye Timber Company log chute in Medio Dia Canyon, on the east side of the Jemez Mountains, 1908. (Courtesy Southwestern Region, US Forest Service)

He makes other things, too; he makes axes and uses them. "Weapon, shapely, naked, wan, from earth's midmost bowels drawn," as Walt Whitman said.

And this broad bench here, where seven and twenty years ago I remember vistas of noble pine, has heard the ringing strokes of the wan weapon since then. The tall pines have gone, the stumps alone are left. No log wagon could be brought up here to draw them down, but the loggers went to the edge of the bench overlooking the valley, and there they built with stakes and trenails a two-thousand-feet-chute of hewn trunks down which their logs slid thundering to the floor below.

Before the days of bulldozers and logging trucks, the only way to get logs off benches and mesas down steep canyons was by hauling them with horses, mules, or oxen to the edge and then sliding them down log chutes.

These were very specialized logging structures, requiring a lot of skill to build and operate. They were dangerous too.

After reading Townshend's story, I found traces of the log chutes below Area 3 on the ground. Along one of these traces were the remains of the long logs ("sticks") laid side by side that served as the chute and short cross beams ("ties") that held them together. I tree-ring dated the outermost rings of the chute's woody remains, finding that it was built in 1884 or 1885.[2] The cut logs that came down the chutes were probably milled here in the valley, and undoubtedly some of the old houses in Jemez Springs today contain boards and beams from them.

In logging terminology, a landing is a place where logs are gathered and sorted before transport to the sawmill. In this case, the logs coming down one of the chutes landed right where the mailboxes are now located on the west side of NM 4, just across from Entrada Lane, the road that goes up to Area 2. The chute ended above the limestone bluffs there, and men tumbled the logs down to the landing. Old, broken-off juniper stumps on those ledges attest to the force of falling logs. Imagine the racket and chaos of logs two to three feet in diameter and ten feet long crashing down there!

During the 1930s and especially after World War II, bulldozers and logging trucks became available, and many roads were carved up the canyon slopes to the benches and onto the mesas. The most valuable trees there were felled and hauled down the roads. Some of these old roads are visible above the Valles Caldera National Preserve offices. The roads leading up to Areas 1, 2, and 3 were originally logging roads. Comparisons of aerial photos from the 1930s and 1950s and recent satellite images on Google Earth show the forest clearing from logging and then forest regrowth. This logging was all on private lands, part of the old San Diego Land Grant. This was not careful forestry and selective cutting, as generally practiced on Forest Service public lands. For the most part, it was "high-grade logging," meaning the largest and straightest stem trees were taken while low-value trees—crooked, forked, leaning, and small—were left behind. In effect, the rule for this style of logging was "cut the best and leave the rest."

Similar cutting took place for decades on the mesas to the west on the San Diego Land Grant, which had been purchased for a pittance from the original Spanish land grant heirs (see chapter 11). In 1910 the corrupt buyers sold timber rights to a logging company with huge profits. By the late 1920s,

Figure 21.2. An ancient, crooked, forked, fire-scarred, and leaning "reject" ponderosa pine tree left on Virgin Mesa following decades of logging.

Chapter 21

railroad logging was in full swing up the Rio Guadalupe and on the mesas above it. To this day, when you travel dirt roads on the benches and mesa tops above the Guadalupe and San Diego Canyons, most of the oldest trees you see are the "leave trees," the "rejects" remaining following the logging of the 1920s to early 1960s.

After the logging, and during wet periods over the past century, large numbers of trees germinated and grew in the understory of the few large trees left behind. Low-intensity surface fires that used to keep the forests open were also suppressed by livestock grazing and active firefighting by the government. This led to very dense thickets of small-diameter, stunted ponderosa pine trees across large parts of the benches and mesas. These thickets burn extremely hot when wildfires occur, "laddering" flames right up into the crowns of any taller trees within them. The Forest Service acquired the mesa lands of the old San Diego Land Grant in the mid-1960s. Since that time, a combination of wildfires, prescribed fires, mechanical thinning of thickets, and some commercial logging has changed the forests on the mesas again.

The thinning of smaller-diameter trees and prescribed burning are having some positive effects now, such as work done on San Juan and Cat Mesas. The Cerro Pelado Fire in April 2022 decreased in intensity when it hit the thinned, prescribed-burned areas there, perhaps saving the Sierra de los Piños neighborhood, which was directly in the fire's path. There is a huge amount of thinning and prescribed burning work still to do, especially on private properties and around homes on the slopes and benches. Relatively few people have taken the initiative to reduce overabundant fuels on their properties, but some have, and the Forest Service has been thinning and pile burning around some neighborhoods. Let's hope this needed work accelerates in coming years.

Last, here is what Townshend said about his hike down from the Area 3 bench in 1903:

> And so, as the sun sloped toward the west, I climbed down parallel to the old chute by the logger's path and drank at a little spring beside

Figure 21.3A–B. The first picture, of tent rocks above Area 3, San Diego Canyon, was taken by R. B. Townshend in 1903. Redondo Peak is in the distance. The recent view (2014) was shifted about thirty feet back and upslope because of in-grown trees obstructing the original view. (1903 photo courtesy Pitt Rivers Museum, University of Oxford)

the trail where they once drank, and where the wild deer now come to quench their thirst, for the fresh deer tracks were everywhere around the spring. I am glad these deer came back and have not given up the game in despair because men once invaded their haunts and scared them away with guns and noise of axes and falling timber and roaring chutes. Not always do the wild things get the chance to return, but when they do, let us be glad and welcome them, and perhaps Pan and the nymphs may come back to us some day.

Part III

Soda Dam, Geology, and Floods

Soda Dam waterfall.

Figure 22.1. This photo was probably taken in the late 1870s. It shows a massive log jam from a previous flood. The whitish log hanging over the main pool got stuck there and remained for about sixty years. The 1941 flood removed it. Notice the cascade travertine dome that formed above the grotto. Below and to the right of the horse and man on the top you can see the old fracture zone, where a great flood breached the dam hundreds of years ago. Photo by Ben Wittick. (Courtesy New Mexico History Museum, Palace of the Governors Photo Archives, NMHM/DCA Negative No. 015574)

CHAPTER 22

Soda Dam, Logs, and Floods

Soda Dam is a fabulous geological and cultural landmark. Technically it is a fissure ridge and cascade travertine deposit, forming a natural dam within a narrow canyon. Fissure ridge travertines are calcium-carbonate spring deposits laid down in a line along a geological fault, a crack in the earth.[1] They are found elsewhere around the world, but no others to my knowledge form a near-complete dam across a river in a narrow canyon. The Jemez Fault crosses San Diego Canyon near Soda Dam, and the central fissure of the dam is aligned along the axis of a secondary fracture of the Jemez Fault.

The Hemish (Jemez) people and their ancestors have lived at and visited Soda Dam since time immemorial. They farmed maize and other crops along the shores of lakes that filled above Soda Dam. Jemez Cave is located directly above the modern Soda Dam within travertine deposits of a much older and larger soda dam that formed more than five hundred thousand years ago.[2]

From 1934 to 1935 archaeologists excavated deep sediments in the floor of Jemez Cave. They found hundreds of artifacts and more than eleven hundred corncob fragments. The oldest corncobs were carbon-14 dated at about three thousand years before the present. This makes Soda Dam one of the earliest known maize agriculture sites in the United States.[3]

There are many interesting stories of geology, climate fluctuations, and people embedded in Soda Dam's travertine and caves. Hopefully, these stories will be told someday in a visitor center or museum and by interpretative signs and guided tours at Soda Dam. One of Soda Dam's stories is about past floods and how flood-uprooted trees played a role in creating the current dam, which is about five thousand years old. This story is revealed in old and recent photographs and by logs stuck in the dam.[4]

Upstream of Soda Dam, the Jemez River flows directly toward the west end of the dam. It makes a hard left turn at the dam and flows toward the east end, where it plunges through a narrow opening under the cascade

Figure 22.2. This photo from 1912 shows the stuck log hanging over the pool. Notice how much the dome over the waterfall and pool has grown since the earlier photo was taken. This dome continues to grow slowly, despite the road blasting through the west end in 1961. Travertine deposits continue here because the top of the dome is below the road level, so some spring flows continue from the central fissure to the west. Photo by Jesse Nusbaum. (Courtesy Palace of the Governors Photo Archives, NMHM/DCA Negative No. 012808)

Figure 22.3. Big floods and the plugging up of Soda Dam's opening have recurred many times, but apparently only twice in the past eighty-plus years (1941 and 1958). The 1941 flood caused a flow event over the top on the east side, as this photo shows. The log stuck above the main pool since about the 1870s can barely be seen through the floodwaters. The hanging log was ripped out during this flood; it no longer shows in post-1941 photos. Also, notice the water pouring into the pool on the west side (*left*), directly below the grotto. This indicates that water was also coming over the top of the dam on the west end, where the old road went over. After the flood, the forked log behind the dam ended up in the main pool below the dome. (Courtesy Adams family)

travertine "dome" that hangs above the waterfall and river. That opening on the east end is only about ten feet wide and ten feet tall. How did Soda Dam force the river to divert its direction of flow toward the east end? And how did the dam form over the top of the river?

One important hydrological fact of Soda Dam is that most of the spring flow volume emerges on the west side of the canyon. Over time the fissure ridge travertine buildup probably proceeded mainly from the west toward the east, gradually pushing the river to the east around the rising fissure ridge deposit. Also, because the opening for the river on the east end narrowed over time, floating logs would catch and get stuck there in the travertine.

When big floods occurred and the rushing waters took down many trees along the banks upstream, the woody debris would plug up the opening. Water then backed up behind the dam, forming a lake, until it spilled over the top. Eventually the log jam in the gap would break apart and the dam opened again. But some logs would get trapped in the travertine deposits continually being laid down by the spring flows coming out of the fractured fissure ridge on the west side of the gap.

On the east end of Soda Dam, you can see a large breach in the nearly level fissure ridge, the result of a massive flood hundreds of years ago. Since that great flood, cascade travertine domes have formed on that end of the dam over the river. They were originally held in place by trapped logs—like wooden rebar in cement. This process of flood breaching, new flood log jams, and travertine rebuilding has recurred countless times over the centuries.

The beautiful grotto on the southeast end near the waterfall and pool is a result of the cascade travertine forming a dome over the top of an open space. Perhaps that space was created naturally by trapped logs above it. Or perhaps people created that original open space. Whether the grotto is natural or human-made, we must recognize the Hemish people's history and interest in Soda Dam.

The cultural, scientific, and educational importance of Soda Dam has been neglected for far too long. The 1961 road blasting through the west end of Soda Dam by the New Mexico Department of Transportation was a tragedy. We taxpayers recently paid millions for a massive new bridge project and a new ranger station just south of Soda Dam. When will we invest in protecting Soda Dam, learning about its many stories, and sharing them with the thousands of curious visitors who stop there every year and wonder: What the heck is this amazing thing?

Figure 22.4. In this 2013 photo, you can see stuck logs on the dam's upstream side, right above the narrow opening. The logs are slowly encased in travertine as spring waters flow over them. However, the reduced flow of spring waters since the road was blasted on the west side has undoubtedly slowed cascade dome formation on the east side. Today, the dome over the narrow opening is precarious, supported by only a few pillars of travertine. There will be future floods and plugging events. When that happens, will water flow over the dam again or will the dome give way and fall into the pool?

Figure 23.1. View from the bridge over the Jemez River on NM 4, just below Soda Dam. On the sloping gray ridge on the horizon are the 500,000- to 280,000-year-old soda dams, formed when the Valles Caldera contained large lakes.

CHAPTER 23

The Many Soda Dams and Lakes of the Valles Caldera

Most people who visit Soda Dam marvel at the strange and oddly beautiful ridge of travertine that crosses the box canyon, with a waterfall at one end and the unfortunate road cut at the other. Few people notice or know that the modern-day Soda Dam is only the most recent travertine dam and the smallest of at least three dams that formed over the past half million years. The remnants of older dams are immediately above the modern Soda Dam, to the west, northeast, and east. They are all visible today as grayish travertine ridges and rocks perched high on the slopes above Soda Dam.

Geologists have used radiometric dating methods to determine the approximate ages of the dams. Core samples were taken from different parts of the travertines with rock drills, and the ratios of different isotopes of uranium and thorium were measured with a mass spectrometer. This method is similar to carbon-14 dating, in that the number of unstable isotopes remaining in the material tested is proportional to how much time has passed since the material was first formed—in this case, the time since the travertine was laid down by the spring flows.[1]

The results of the dating work are that the oldest and largest travertine deposit, located above Soda Dam to the west, is about 280,000 to 500,000 years old. Most of the upper part dates from about four hundred thousand to five hundred thousand years ago. The other older dams, located above and to the northeast and east of Soda Dam, are about fifty thousand to one hundred thousand years old. The modern Soda Dam is a baby in comparison at less than ten thousand years old, and most of it was likely deposited in the past five thousand years.[2]

Figure 23.2. The center of this photo, taken around 1900, shows the dam (fifty thousand to one hundred thousand years old) on the north side of Soda Dam and the Jemez River. (Courtesy New Mexico History Museum, Palace of the Governors Photo Archives, NMHM/DCA Negative No. 057902)

In addition to their great age, the oldest dams are massive. The oldest set of deposits, on the west side, is more than one hundred times larger in volume than the modern Soda Dam. Incidentally, Jemez Cave is located immediately beneath these largest and oldest travertine deposits. You can see this dam as you drive on NM 4 north of Jemez Springs, especially as you cross the bridge over the Jemez River right by the new Forest Service ranger station. When you look up at that huge wall of travertine directly in front of and above you, imagine the enormous quantity of hot spring flows that must have occurred to create that deposit over thousands of years.

The next oldest dam is located right behind Soda Dam as you view it from the south side driving up NM 4. All of these old dams were formed as a consequence of hot mineral waters flowing down through the Jemez Fault zone and related fissures. The Jemez Fault is a big crack in the Earth, and it extends on a northeast-southwest line across the southern rim of the Valles Caldera. The Jemez Fault crosses San Diego Canyon right at Soda

Dam. The dams are fissure ridge travertine deposits, created by hot spring mineral deposition laid down along the linear fissures.

The travertine is primarily calcium carbonate. It comes from the hot waters traveling down through the Madera limestone, which is visible as yellowish and gray rocks between Soda Dam and Battleship Rock. The Madera limestone was created from coral reefs and shellfish in marine environments about 300 million years ago. The hot bicarbonate waters emerging at Soda Dam, with temperatures of about 47 degrees C (116 degrees F), originate from the very hot (300 degrees C [572 degrees F]) and deep (more than 430 meters [1,410 feet]) geothermal reservoir in the Valles Caldera.[3] These hot waters dissolve the calcium carbonates of the old limestone, which are then redeposited and emerge along fissures at Soda Dam. The travertine is essentially young terrestrial limestone that originated in old marine limestone.

Here's where the geological story gets more interesting, and a bit mysterious. Given the different ages and sizes of the multiple travertine dams, two questions arise: Why did these distinctly different dams form when they did? Why are the oldest dams so large in comparison to the modern Soda Dam? The partial answers to these questions are provided by a theory described by the foremost geologist of the Jemez Mountains, Fraser Goff, and his colleagues at Los Alamos National Laboratory.[4] Some readers may remember one of those colleagues, the late Jamie Gardner, who lived in Jemez Springs for some years. Here, in a nutshell, is my summary of their very interesting theory:

Hot magma—molten rock—is not very far beneath the floors of the valles within the caldera.[5] The heat from that magma causes very ancient waters from Earth's crust to rise upward within the caldera. At the same time very old geothermal waters are driven upward by the heat, vast amounts of modern rain and snowfall are captured within the giant crater that is the Valles Caldera. These waters coming down from the sky combine with the waters rising from the deep. Then they flow downhill as waters tend to do. The floors of the valles are around eighty-five hundred feet in elevation, while Soda Dam and Jemez Springs are at around sixty-three hundred feet. That is a big drop, and as water pipeline managers know, such differences create a hydraulic head—that is, water flowing under increasing pressure as it drops to lower elevations.

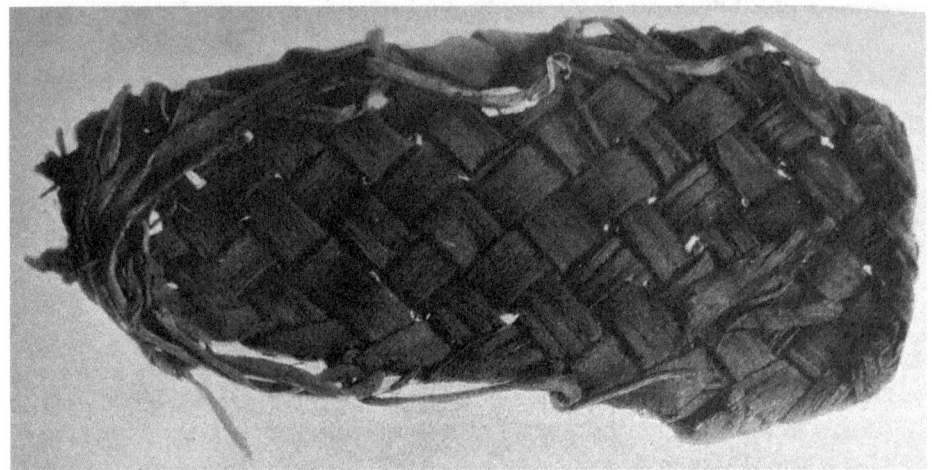

Figure 23.3. (*above*) View to the west, from upstream of Soda Dam. The oldest and largest travertine dam is in the center, with Jemez Cave located beneath it. The next oldest dam is on the right, and the central fissure is just visible through the trees. The top of Soda Dam is visible at the bottom of the photo.

Figure 23.4. (*below*) A yucca fiber sandal from Jemez Cave, one of many found there (From Alexander and Reiter, *Report on the Excavation of Jemez Cave*)

Chapter 23

The waters emerge along the Jemez Fault at Soda Dam as artesian flows that rise above the local level of the ground and creek bed. Notice how level and high the top of Soda Dam is relative to the current canyon floor. This is from the artesian flow of hot waters and the travertine deposition along the fissure over the past five thousand–plus years. So again, why did the massive spring flows and travertine deposits create large fissure ridges during different time periods?

Goff and his colleagues propose that the big fissure ridge travertine dams were formed when there were large lakes impounded in the Valles Caldera. The oldest and largest travertine dam built up after lava flow eruptions from South Mountain and San Antonio Mountain blocked the East Fork and San Antonio drainages. Eventually these lava dams eroded away and the lakes drained out.

Then there seems to be a hiatus in massive travertine deposition until after about one hundred thousand years ago, when multiple eruptions of the youngest Valles Caldera lavas and tufts (from about sixty-eight thousand to seventy-four thousand years ago) periodically clogged the drainage of the East Fork. These volcanic eruptions consist of (youngest to oldest) Banco Bonito lava and the underlying El Cajete–Battleship Rock pyroclastic deposits. Battleship Rock was formed by one of these volcanic deposits.[6] Eventually all the volcanic dams eroded away, and catastrophic floods roared down San Diego Canyon.

During the times when large lakes were impounded within the caldera there was greatly increased flow of meteoric and hydrothermal waters out the south side of the caldera, down through faults and fissures in the limestone in San Diego Canyon. There the waters arose as hot springs and travertines where the Jemez Fault crosses the canyon. In sum, large lakes in the caldera were responsible for the large soda dam formations, and hiatuses between the big formations occurred during periods when the lakes had drained out.

If this theory is correct—and much of the geological evidence that Goff and others have found supports it—why did the modern Soda Dam form in the past ten thousand years? There have been no large lake stands in the valles during this period to our knowledge. This is a mystery yet to be solved, but there are several possible explanations. Perhaps climate and vegetation variations over the past ten thousand years led to increased spring flows, but not in the great amounts that occurred during large lake stands in the valles. Or pulses in heat arising from the deep magma chamber may have increased hydrothermal flows during this period. It is also possible that there were many small dams like the modern Soda Dam in the past, but they tended to erode away or get flushed out occasionally by giant floods. There is not a single or simple explanation for episodic formations of travertine fissure ridges in San Diego Canyon.

All in all, these deep time perspectives of the amazing volcanic, hydrological, and travertine histories of the Valles Caldera serve to remind us that the only constant in this dynamic landscape is change itself. From our modern and personal perspectives, it is tragic that the west end of Soda Dam was blasted in 1961 for the convenience of building NM 4. But from a longer perspective, we can be assured that another soda dam will arise here in centuries to come.

CHAPTER 24

The Blasting of Soda Dam

Sometime in early 1961, the western end of Soda Dam was dynamited and bulldozed to level the roadway for the paving of NM 4. The New Mexico Highway Department apparently had no second thoughts about the wisdom of blasting a world-class geological and cultural landmark of the Jemez. No record of the decision process has come to light, only letters from the private landowners of Soda Dam to the highway department, confirming their donation of the easement and their gratitude for improvement of the road for Jemez Valley residents and visitors.[1]

The steep and rocky climb over Soda Dam had been a bottleneck on the road for decades. It was a difficult passageway, especially for freight wagons and later for big trucks. Many people probably welcomed the leveled and paved road. The improved highway was used in promotions for the sale of summer home sites in the "Jemez Country" development, known today as Areas 1, 2, and 3.[2] However, some old-timers in the valley told me that they immediately lamented the damage done to Soda Dam. Early on it was clear that the blasting on the west end had diverted the flow of spring waters out of the dam, causing reduction or cessation of spring flows over the eastern end above the river.

The large gap blasted in the dam exposed major channels of spring waters that formerly flowed along and up through the central fissure. During the wettest eras of the past, those waters would rise up and flow over the top of the dam, forming small travertine cones in several places where spring flows were concentrated. After the blasting, those waters and the travertine had to go somewhere else.[3]

The original engineering plans for the blasting show that the road builders were aware they would have to redirect the emerging spring waters coming up in the gap. In addition to the removal of the entire west end of the dam, they also dug a trench below road level, one hundred feet long by

Figure 24.1A–B. In a photo taken by R. B. Townshend in 1903 a freight wagon pulled by six horses ascends the south side of Soda Dam. The same view in 2017 from the top of Soda Dam looking south. Note the cattle guard over the channel under the road, which has now been removed. (1903 photo courtesy Pitt Rivers Museum, University of Oxford, Catalog No. 1998.58.87)

Chapter 24

twenty feet wide by two feet deep, and filled it with rocks four to six inches in diameter to create a so-called French drain.[4]

The drain helped stop spring waters from bubbling up through the roadway, but it did not solve the problem of waters emerging from the west cut bank about five feet above road level. The road engineers had to install a channel under the road with a heavy-duty cattle guard over it for those waters to flow to the east side of the road. Some waters still flowed down the west side of the road. Every few years they had to lift off the cattle guard and ream out the accumulated travertine in the channel, as well as down along the west side.

Most recently, the highway department removed the cattle guard and filled the channel carrying water to the east side. Using a backhoe, highway workers also ripped out the travertine mound built up where the spring emerges on the west cut bank and the little pools visitors loved to soak in—within arm's reach of passing traffic! All the spring waters from the cut bank are now flowing down the west side of the road and reaccumulating travertine there.

The west end of Soda Dam is like a bleeding wound that partly scabs over but is never allowed to heal. The lack of channeled spring flows from the west side has resulted in low or no accumulation of travertine on the main dome over the river during dry years. During the recent drought, with low spring flows, the dome took on a dusty brown color. In the early 2020s, with better rains, the hot spring flows out of the central fissure increased somewhat and the dome over the river turned the whitish color of the newly deposited travertine.

That's a good thing because the main dome is attached to the Precambrian granite on the east side of the river by relatively small travertine pillars. If those pillars are not reinforced by more accumulated travertine, it is only a matter of time before the dome collapses into the river. Floods and log jams on the upstream side could hasten a collapse (see chapter 23).

Concern about a crumbling Soda Dam goes back to the early years following the blasting. The Jemez Mountains Wildlife Federation recognized the damage done to the dam and sought protection and restoration in public ownership. They began a long-term campaign in 1967 by asking the private owners to donate the dam to the Forest Service.

Figure 24.2A–B. (*opposite*) The most difficult part of the old road up and over the west end was near the rocky top. Note that two men are steadying the horses on each side. The equipment on the freight wagon appears to be the top of a steam boiler. It was probably a replacement for the boiler that exploded in May 1903 at the sulphur mine northeast of present-day La Cueva). (Early photo courtesy Pitt Rivers Museum, University of Oxford)

Figure 24.3. (*above*) In this photo, taken on June 11, 2023, the central fissure in the blasted end of Soda Dam is visible just above and to the left of the car. The spring flows emerging from the west cut bank are visible on the right side of NM 4.

In a 1969 newspaper article, Leonard Tartaglia, a member of the federation, said, "Residents of the village do request that everything be done to convert or re-channel the wasted hot springs back to its original and natural course to supply nature's food for survival of Soda Dam." He continued, "Destructiveness does not bring prosperity to our state monuments. We do beg the general public and our news media to share our protests and inducements to revive the once renowned state monument."[5]

In announcing the Forest Service's full acquisition of the dam in 1977, the New Mexico Wildlife Federation newsletter recounted the history of the campaign:[6]

> The dam, until now, has been in private ownership, thus preventing any public effort to preserve the structure, which has been crumbling since the highway department damaged it about 25 years ago.
>
> It took three separate donations to accomplish Forest Service ownership. In 1969 Calvin and H. B. Horn donated the downstream face of the structure. In 1976, A. E. (Tommy) Thomas donated over 250 acres which included both sides of the stream, just below the dam. About seven years ago Carl Seligman committed to donate the upper half of the dam. Unfortunately, he died before the donation could be made and it has taken this long for the heirs and the Forest Service to effect the final donation. . . .
>
> The issue remained dormant with little progress being made until summer of 1974. Catherine Stephenson, who had been one of the earliest advocates for public ownership of Soda Dam, was then president of the Jemez Mountains Wildlife Federation. She began writing the Seligmans on behalf of the Jemez chapter in an effort to renew their interest in donating the dam. In January of 1975, the Seligmans communicated their interest to donate their side of the dam.

From the beginning, the Forest Service was uncertain how to manage Soda Dam, although the Jemez District ranger at the time, Charles McGlothin, said, "The prime concern to the Forest Service is providing for public safety, and protection and preservation of the Soda Dam." Further, he said the agency was developing a master plan for accommodating tourists at the site while minimizing destruction of the dam itself. The ranger asked for suggestions from the public.[7]

The Forest Service may have produced a master plan in the late 1970s and implemented it, but, if so, the plan was minimal. To this day there are no interpretative signs describing the geological and cultural history of Soda Dam, and people continue to climb and walk on the dam and the domes over the river as much as ever. The managers closed Jemez Cave to the public some years ago after it was discovered that rock climbers had driven numerous pitons into the overhanging roof.[8] Jemez Cave, of course, is a key part of the Soda Dam geological and cultural landscape and is also a world-class archaeological landmark in the Jemez. But it is now practically invisible to the public.

In 2021 Assistant Professor Anthony Fettes tasked students in his landscape architecture class at the University of New Mexico to create a set of possible designs for public viewing of and learning about Soda Dam. The designs included viewing decks near the road, interpretive signage or kiosks, trails and bridges over the river and around the dam, railings and boardwalks where needed to keep people off the crumbling parts, and a trail up to Jemez Cave allowing visitors to look inside and learn about its history but not enter. Several of the students called for replumbing the hot mineral waters from the west side back into the central fissure, with pumps driven by solar power if necessary.

Clearly, it is long past time for Soda Dam to be managed and restored, protecting its present and future while recognizing and teaching about its amazing geological and cultural history. The Forest Service should step up to the plate to finally accomplish what it promised more than forty-six years ago. Alternatively, perhaps Soda Dam and Jemez Cave should be transferred to the Valles Caldera National Preserve, National Park Service, or New Mexico State Parks or Historic Sites and comanaged with the Jemez Tribe. Soda Dam is certainly an important and extremely interesting part of the Valles Caldera geological story and of Hemish cultural history and traditions. In any case, agency leadership and a serious investment are needed by the state and federal government to fully protect and appropriately share this special place with everyone.

CHAPTER 25

The Caves of Soda Dam

Among the many wonders of Soda Dam are the caves found beneath the ridges and domes of travertine blocking and spanning the Jemez River. Two of these caves are visible today, but another large cave is now buried and out of sight. Most visitors to Soda Dam see the beautiful grotto on the south side of the dam, immediately west of the waterfall. Some people look upward to the west side of the canyon and see the mouth of Jemez Cave. This large cave is beneath the massive travertine deposits of the oldest and largest dam, which formed about 280,000 to 500,000 years ago.

The small grotto on the south side of the dam, so named because of its size and possible origin, is very wet, with active spring flows out of the opening. A single pillar bisects the entrance, and a beautiful travertine basin overflows with spring waters in the back corner. Blue and green algae grow on the rippled travertine ceiling and walls.

The visible caves are interesting enough, but almost no one is aware that there is a large hidden cave under the north side of Soda Dam and that it was accessible during the late 1800s and mid-1930s. I first learned of this cave when I read an account by J. K. Livingston of his family's horseback camping trip to Jemez Springs in the summer of 1886 (see chapter 7). This excerpt is from Livingston's article in the *Santa Fe New Mexican* in July of that year, describing Soda Dam, where his family camped:[1]

> It is an absolute fact, that within one hundred feet of camp there were not less than forty-seven distinct hot springs, and that each member of the party, even to the 2-years old baby, had a private bath tub, chiseled by nature in the rocks, and filled at all times with sparkling soda or sulphur waters, of just the right temperature for bathing; that we frequently took natural vapor baths, by visiting a certain grotto on the *north side of the dam* [emphasis added], within the walls of which we

also enjoyed copious morning draughts of a most delicious hot mineral water, very similar in action, if not in taste, to that of Saratoga's most popular spring, the "Hawthorn," where the waters of the latter are heated artificially, and so furnished visitors when desired; and lastly, that twelve feet above our canvass domicile, and easily reached by steps cut out of the rock, we fitted up a library and reception room, or rather, nature fitted it up, in the solid rock; of dimensions thirty feet long, ten wide and thirteen feet high; with an entrance twelve feet wide, and a window (overlooking the cataract) three by four feet, whose embellishment of halls ceiling and pillared portico, were in pure bi-carbonite of soda, with some adulteration of lime and alum-after the fashion of all baking powders, if we are to believe certain rival advertisements!

Today, there are no visible caves on the north side of Soda Dam that match this description. Another mention of this cave is in a report on the archaeological excavation of Jemez Cave from 1934 to 1935 by H. G. Alexander in *El Palacio* magazine:[2] "It is peculiar that the upper cave [Jemez Cave] should ever have been used when there are caves so well situated closer to the water supply, warmed by the hot mineral springs which have formed them, and also as well protected from the weather. They are used throughout the winter by our workmen as ideal camping sites."

The workmen Alexander refers to are the men who helped excavate Jemez Cave throughout the winter of 1934 to 1935. Remarkably, the caves under the north side of Soda Dam were "ideal camping sites"—warm, dry, and with enough room to accommodate wintertime shelter for four men.[3] Again, the smaller and very wet grotto on the south side does not fit this description.

Alexander explains that Jemez Cave seemed to have been utilized by the Hemish people until a specific time in the remote past. He suggests that the Hemish might have shifted their ceremonial activities to the lower "caves" (note the plural) below Soda Dam when they were formed and became accessible. He mentions that the archaeologists were informed about a Jemez legend that describes this transition. In response to a request from an elder Hemish man who visited them, the archaeologists refrained from excavating the lower caves.

From the historical descriptions and likely location of the north-side cave, I suspect that it was created, buried, and uncovered multiple times during the past five thousand years, which geologists estimate is the age of the modern Soda Dam. Here is a theory about the hidden cave (or caves):

Upstream of Soda Dam, the Jemez River heads directly south toward the far western end of Soda Dam. At the base of the dam the river makes a hard left turn toward the east and flows down the length of the dam, where the waterfall plunges under the domes on the eastern end. Many centuries ago a massive flood broke off and washed away the travertine over the eastern end of the dam. This break in the dam is visible today as a zone of sheared-off travertine on the central fissure ridge just west of the cascade domes. The cascade domes over the river formed since this great flood.

During that flood, or a subsequent one when the dam could not hold back waters or sediment, the travertine ridge on the western end of Soda Dam was undermined by the fast-moving waters, exposing a large cave beneath the dam on the north side. That cave was apparently open and accessible during the late 1800s and mid-1930s, until the floods of the twentieth century. The 1941 floodwaters, which backed up behind and overtopped Soda Dam, deposited large amounts of sediment on the north side, effectively burying the large cave. Then, in 1961, the western end of the dam was dynamited to level and pave the road. Some of the debris was bulldozed over onto the north side, completely covering the entrance of the large cave (see chapter 24).

Similarly, Jemez Cave may have originally been carved out by the ancestral Jemez River when it was about one hundred feet above the current river level. Jemez Cave is essentially a deep cavity with an overhanging roof beneath the largest and oldest travertine deposit. Thousands of years ago, a flooding Jemez River probably undermined that dam, creating Jemez Cave.

Given this history, it is not surprising that stories have long been told about a hidden cave under Soda Dam. One of those stories is recounted by Roland Pettit in his 1975 book, *Exploring the Jemez Country*:[4] "Not only is Soda dam a geological wonder, it was also an archeological storehouse. For many years, stories were told of a dry cave under the dam, accessible only by an underwater passage. One of the valley's early settlers, Moses Abousleman, successfully found the entrance and retrieved many Indian artifacts from it, including several handwoven blankets."

Moses Abousleman died in 1934. This suggests that if the story is true, and the cave was the same one used by workmen from 1934 to 1935, then

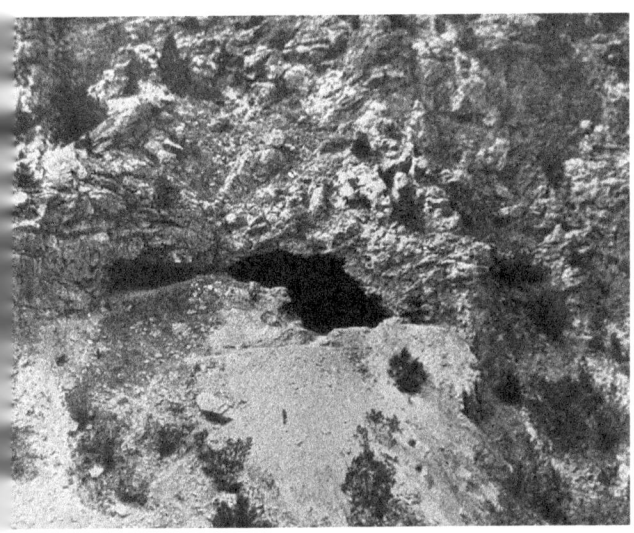

Figure 25.1A–B. Jemez Cave in 1935 after excavation by Alexander and Reiter and corn cob fragments collected from the southeastern area of Jemez Cave by archaeologists in 2013. (Cave photo from Alexander and Reiter, *Report on Excavation of Jemez Cave*; corn cob image from Adler and Stokely, *Jemez Cave Condition Assessment*)

the main entrance had previously been buried when Abousleman found his way into it via "an underwater passage" some decades before.

Finally, although many photographs of Soda Dam were taken before the 1930s, I have yet to find one that shows the north side of the dam. Visitors with cameras are most impressed by the beautiful and strange forms on the south side with the long travertine ridge, pretty grotto, domes, and waterfall. Someday, if a pre-1940s photo of the north side shows up, we just might see the hidden cave.

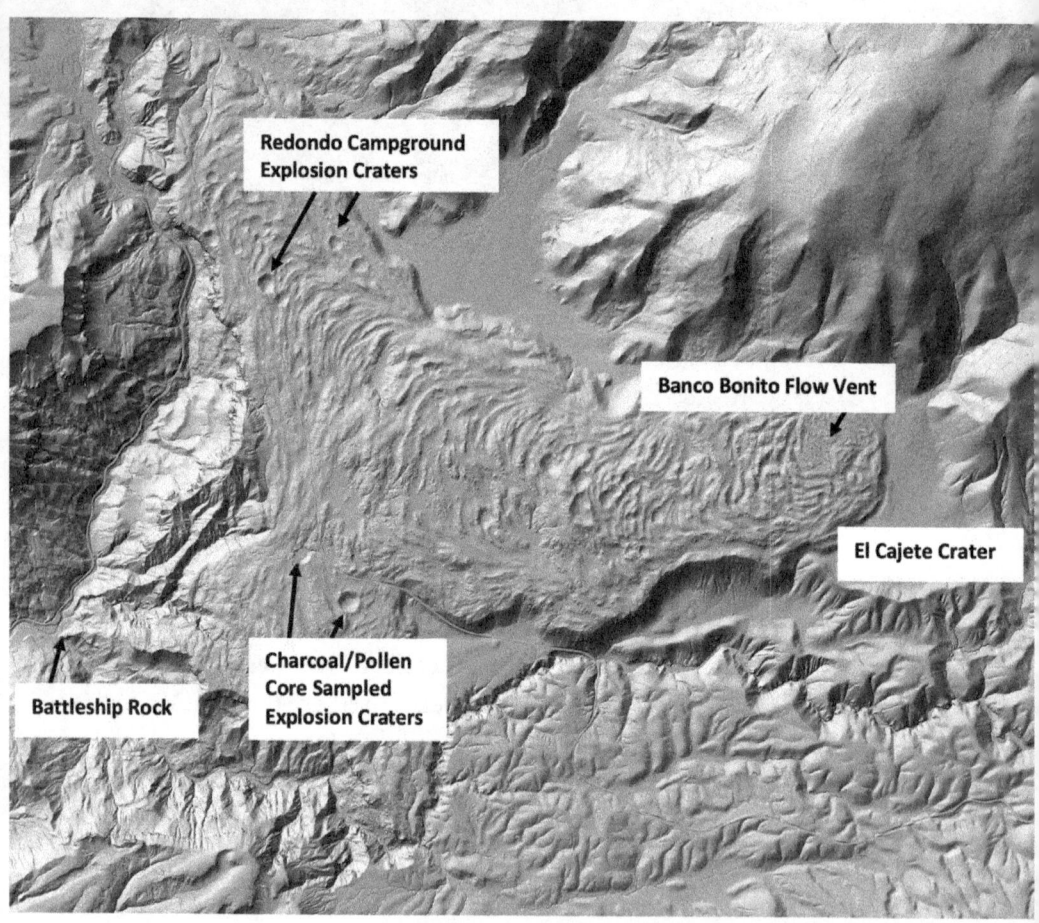

Figure 26.1. Lidar map of Banco Bonito showing pressure ridges and explosion craters, 2012.

CHAPTER 26

The Explosion Craters of Banco Bonito

If you have ever camped or picnicked at Redondo Campground and wandered around that hilly, rolling landscape, you may have noticed the explosion craters there. These circular pits were created by steam blast (phreatic) explosions through rhyolitic lava flows that erupted from the Banco Bonito vent. Banco Bonito consists of two main flows that date back sixty-eight thousand to seventy-four thousand years ago.[1] The surface of the Banco Bonito flows is traversed by NM 4 from a point roughly a half mile west of Jemez Falls Road to about a quarter mile northwest of Redondo Campground. The Banco Bonito lavas, Battleship Rock ignimbrite, and South Mountain lavas blocked the East Fork and San Antonio Rivers multiple times in the past, creating temporary lakes in the valles.[2]

The explosion craters are most visible in airborne lidar (light detection and ranging) scans of the Jemez Mountains obtained in 2012.[3] The lidar data are used to create a three-dimensional map of the ground surface. The data are obtained by bouncing pulses of laser beams off points on the ground back to a sensor on an aircraft. The laser device and sensor, along with a very high-resolution geographic positioning system and a lot of computer work, result in very precise x, y, and z coordinates of the points on the ground. Millions of these points are then stitched together to create a high-resolution 3D model (or map) of the ground surface.[4]

The ground surface map of Banco Bonito clearly shows the flow patterns of lava (arcuate pressure ridges) as it emerged from the vent on the southeast flank of Redondo Peak. On this map Banco Bonito looks like an elongated puddle of molasses that flowed from the northeast to the southwest, toward San Antonio Creek and the East Fork. Wave-like pressure ridges and

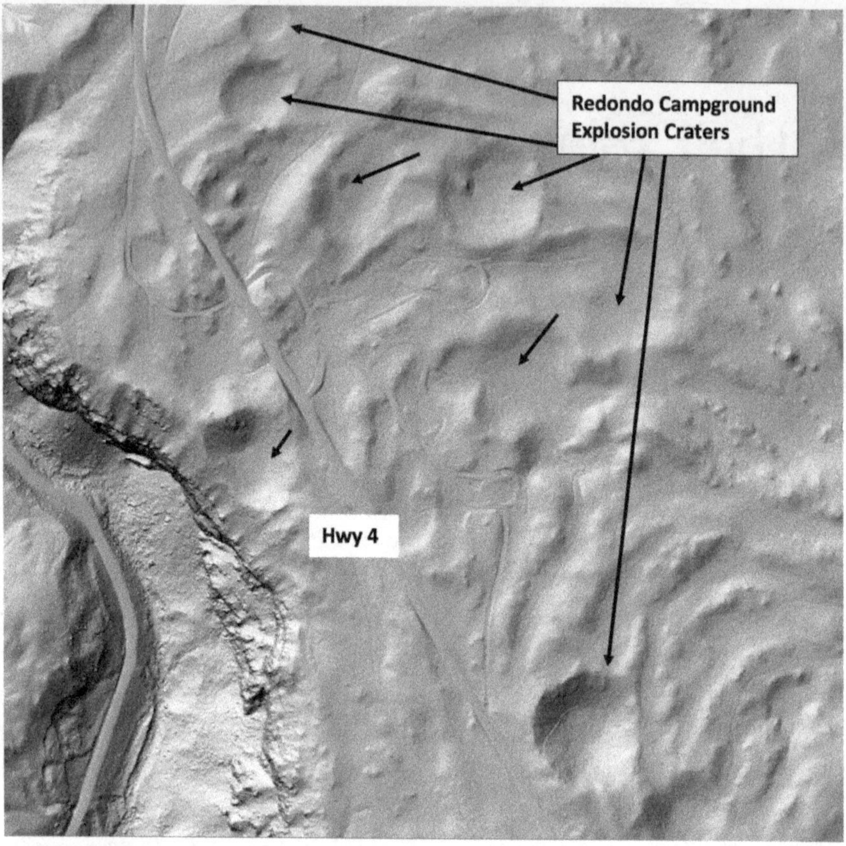

swales were formed as the viscous lava flowed out of the vent, with circular indentations marking the explosion craters.

Scientists have studied the explosion craters in Banco Bonito and similar lava craters elsewhere. The explanation for their origin is a bit complicated, but the general theory is like this:[5] As lava flows it tends to form layers, with cooled and rigid layers at the top. In the case of Banco Bonito, the flows covered a preexisting layer of pumice, which was formed from volcanic ash. The El Cajete and Battleship Rock eruptions around seventy-four thousand years ago laid down thick layers of pumice and ignimbrite (pyroclastic fall and flow deposits). These units are relatively soft; thus, erosion carved shallow canyons and ravines into the pumice and ash that was next partially filled with gravels, sands, and soils.[6]

Pumice is typically a very lightweight rock and is full of holes. The very porous pumice layers accumulated water from rain and snow. Then, about sixty-nine thousand years ago, hot lava flows overrode the El Cajete–Battleship Rock layers and somewhat younger gravels, vaporizing the water and creating pressure beneath the lava surface. The pressurized steam from heated water in the pumice, and volatile gases from the lava, accumulated into large pockets, or bubbles, and pushed up through the more rigid lava layers above. When the pressure got high enough, the bubbles violently exploded through the upper lava layers, creating the craters.[7]

For many thousands of years, the explosion craters have served as natural traps for dust, charcoal, pollen, and other fossilized plant parts that have blown or washed into them. My colleague Chris Roos, a geo-archaeologist at Southern Methodist University, took sediment core samples more than fifteen feet deep from the bottom of two of the largest craters. He has been studying the charcoal and plant fossils in the core samples to learn about fire, vegetation, and human history in the Jemez Mountains.[8]

Figure 26.2. (*opposite above*) Lidar map of explosion craters around and within Redondo Campground. NM 4 and the loop roads within the campground are also visible.

Figure 26.3. (*opposite below*) Chris Roos and colleagues extract an eighteen-foot-long core sample from the floor of an explosion crater on Banco Bonito. (From Roos, "Long-Term Context")

Carbon-14 dating showed that the record of fire and vegetation from one crater extends back more than twenty-five thousand years, and more than six thousand years in the other crater. One of the most significant observations from counts of charcoal abundance through the length of the sediment cores was that the greatest accumulation rate occurred during the past two thousand years, especially since about 1200 CE. This approximately corresponds with the arrival and major population increase of the Hemish people on this landscape. We interpret this finding as indicating the extensive use of low-severity fires by Hemish people to manage forest fuels and for agricultural and other purposes.[9]

It is remarkable to contemplate that ancient gaseous explosions in lava flows here in our backyard created these large circular pits and that over the past millennia they have gathered bits and pieces of plants, burned forests, and other microscopic relics of the environment. With modern technology and tools, we can map these pits in detail and study the tiny treasures they have preserved in their sediments. New microscopy, DNA analysis, and other measurements are enabling scientists to learn what plants lived here over time and how humans changed things as their populations increased and decreased over the millennia.

CHAPTER 27

Great Floods of the Rio Jemez

Anyone who has lived in the Jemez Valley for some years knows that the Jemez River and its tributaries run muddy and high at times, especially in the spring. The oldest residents of the valley recall the great floods of 1941 and 1958, when multiple bridges, roads, and buildings were washed away. In April 2023 we wondered if another epic flood was coming as an above-average snowpack in our mountains melted out quickly. Ultimately, the 2023 peak flow (on April 13, at 1,310 cubic feet per second) was the highest since 1987 and the eighth largest since 1941.

From Jemez River gauge data measured near Cañon, it is clear that 1958 was the largest flood year in this record (back to 1953). However, the 1941 flood was probably larger. I say *probably* because we don't know precisely how extreme that flood on the Jemez River was because the gauge station was washed out. The record in the graph covers only the period after the gauge was replaced.[1]

Newspaper articles and historical studies offer insight into the largest flood years of the Jemez River and its tributaries. The oldest floods are best known along the Rio Grande, and although these would not always correspond with floods on the Jemez, many of the largest ones probably did. For example, a May 27, 1948, newspaper article lists some especially large floods that may also have occurred on the Rio Jemez:[2] "History records numerous floods since white settlers first entered this area. The first being in 1828 at Tome and El Paso. Old newspapers give accounts of floods in 1866, 1874 and 1884. The 1874 flood was the worst in recorded history. US Engineers estimate the peak flow was about 100,000 cubic feet per second—more than three times the flow of the bad 1941 flood. The 1874 flood is referred to as a '100-year flood.'"

In an extensive review of flood events using Spanish-, Mexican-, and American-era documents, historian Dan Scurlock listed dozens of floods on the middle and upper Rio Grande, including a few that he called

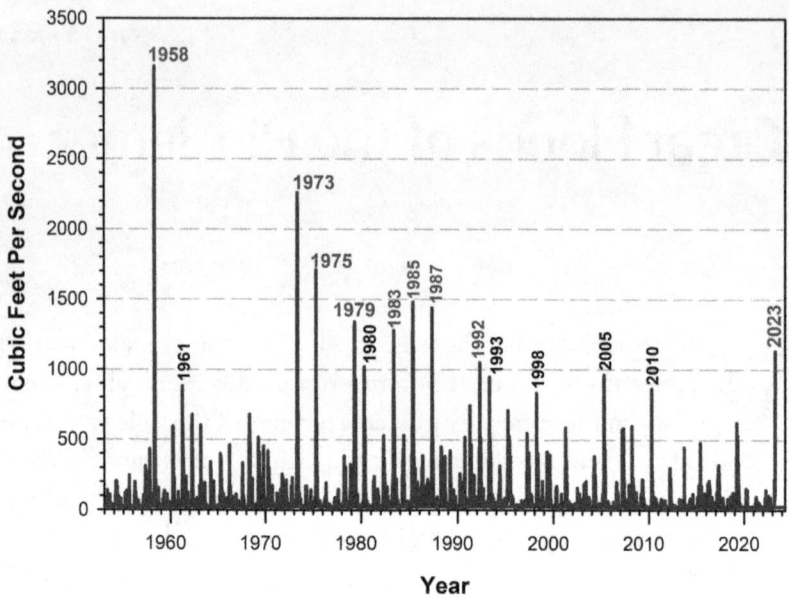

Figure 27.1. Average daily flows at the Jemez River gauge in Cañon, 1953 to 2023.

megafloods—those exceeding one hundred thousand cubic feet per second (cfs) in 1828, 1872, and 1884. He estimated the 1874 flood at forty thousand cfs. He doesn't provide flow estimates for the 1941 flood (because of the loss of gauges?), but it seems to have approached the size of the 1874 flood. Scurlock also said this about the 1941 flood:[3] "Twenty-nine inches of precipitation fell during this period [January to May]; widespread property damage; more than 50,000 acres inundated in [the] valley."

I don't know which of these earlier megafloods or very large floods also roared down the Rio Jemez and its tributaries, but it's clear that the 1941 flood was probably the largest during the twentieth century. However, photographic and documentary evidence shows that huge floods in the late nineteenth century were much larger than anything seen in the twentieth century. They ripped trees out along the Jemez River bottom and tore up roads and bridges. This is evident in photos of Soda Dam from the late 1800s showing huge logjams on the far eastern end. The 1941 flood backed up behind Soda Dam and overtopped the far eastern and western ends, but it apparently did not leave a large log jam.

Figure 27.2. Log jam from a very large flood on the eastern end of Soda Dam. This photo is undated, but based on the size of the dome over the river compared to later well-dated photos, it is probably from the 1870s. Perhaps this log jam is from 1872 or 1874, when very large floods were documented along the Rio Grande. (Courtesy New Mexico History Museum, Palace of the Governors Photo Archives, NMHM/DCA Negative No. 142680)

Soda Dam and the box canyon immediately below it are spectacular places to view the river in flood stage. I witnessed the 1973 spring flood there, the second largest since 1941. I recall standing in front of the grotto at Soda Dam and watching the roaring waterfall. If memory serves, there was a small amount of water coming over the far eastern side below the dome. What I remember most vividly was that, in addition to the sound of the waterfall, there was a low rumbling noise and vibration in the ground—it was the sound and shaking of large boulders being tumbled along the riverbed by the torrent.

Before the bridges were built below Soda Dam, the crossing could be perilous in flood stage, as this 1915 article relates:[4]

Figure 27.3A–B. The bridge that washed out in 1941 in an undated photo, probably from the 1920s or 1930s. Soda Dam is visible in the distance. Notice the three kids playing on the bridge and some cows just beyond it. The new photo was taken with a drone on July 28, 2023. The old concrete bridge abutment on the east side of the river is visible through the willows at bottom center. (Early photo courtesy New Mexico History Museum, Palace of the Governors Photo Archives, NMHM/DCA Negative No. 057903)

Chapter 27

Jemez Springs, N.M., April 15—The Jemez River is on a rampage. For about a week, the water has risen a little higher each night and last night was very high. At the Jemez Indian Pueblo, the river has spread out into the fields.

Harry and George Fluke narrowly escaped a serious accident here yesterday afternoon. When fording the river near the soda dam with four mules and a loaded wagon, a wheel broke and Harry was thrown into the icy water. He was carried about 50 yards downstream in the raging torrent, but finally managed to swim ashore. They succeeded in saving everything.

This afternoon, a more serious accident occurred. A ranchman from the Cueva, starting home from here with four horses and a load of supplies, lost three of the horses by drowning. The young man is prostrated tonight at a nearby house. The horses were pets and he was unable to do anything but sit in the wagon in the middle of the stream and see them die.

Fortunately, Harry Fluke came along and soon got Mexicans and Indians to rescue the young man from his perilous position and, piece by piece, to save the cargo and later the wagon. There has long been a crying need for a bridge at this point.

The first bridge was built sometime between 1915 and the 1930s. This bridge was washed out by the massive May 1941 flood. Gilbert Sandoval, who lives just downstream from the existing NM 4 bridge, recalls as a boy watching the remains of this old bridge floating downstream and crashing onto the bridge that crosses the river to the Sandoval home at that point. Amazingly, the Sandoval bridge held. Gilbert also told me that he and his bride, Genevieve, were on their honeymoon in spring 1958 but were unable to return home for some days because NM 4 had washed out at several places below Jemez Springs.

The 1958 flood isolated Jemez Springs for many days while road crews worked to reopen the washed-out roads and to replace bridges. This newspaper account from April 22 reflects the havoc during the first days of that flood:[5] "Jemez Springs—This little mountain town was isolated Monday night by heavy floodwaters which ripped out bridges, culverts and telephone poles. Telephone company repairman were trapped in an area where a bridge

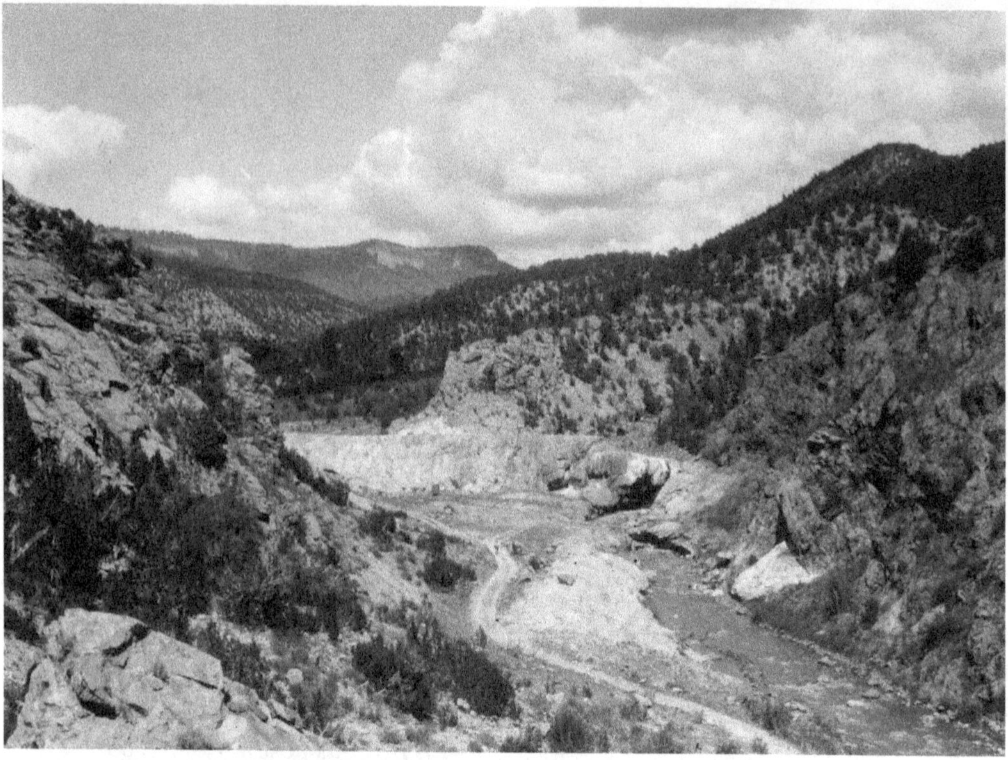

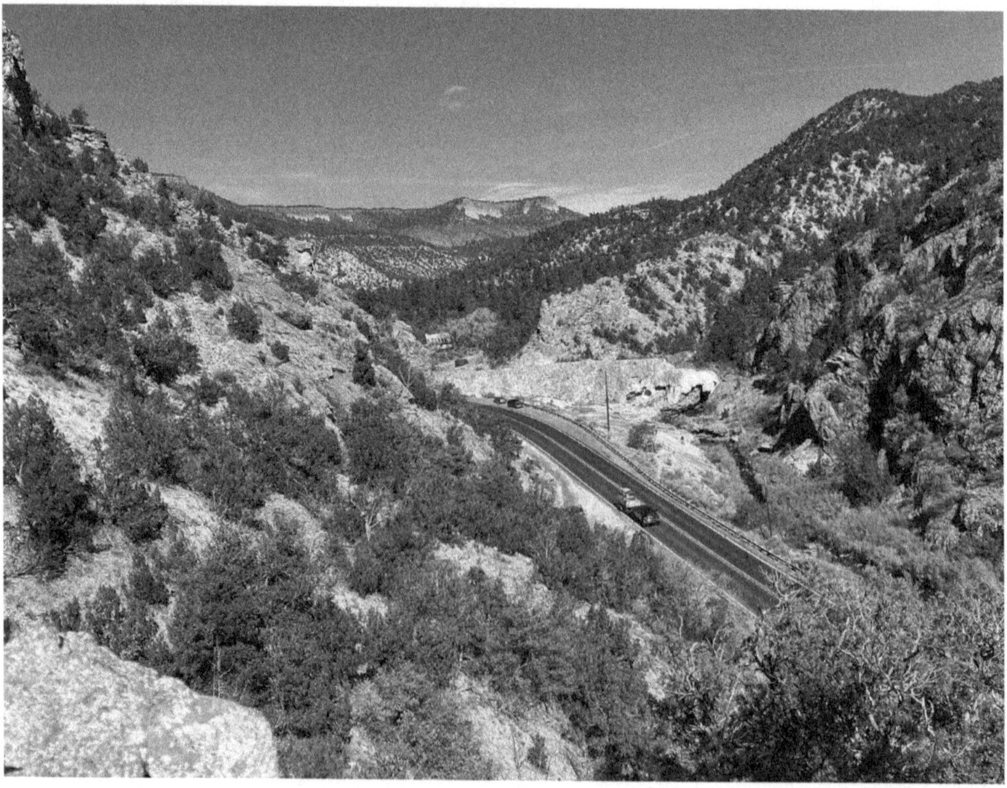

Chapter 27

went out. They reported over an emergency line that the Jemez River, normally about 10 feet wide, was running 200 feet across South of here. Water was running a foot deep across State Road 4. . . . Bridges on State Road Four were out both South and north of here, and big culverts, which normally bridged the stream at La Cueva, also were ripped out."

An article from April 25, 1958, included photos of a washed-out section of NM 4 and a damaged structure that appears to be the old building on today's Bohdi Manda compound. That building was originally the Stone Hotel built by the Oteros, which had a west wing that extended almost to the river. The 1958 flood apparently took out most of that wing.[6]

Jemez Valley residents today should bear in mind these great floods of the past, especially folks living near the river! Since the 1941 and 1958 floods, the river channel was stabilized at some points with boulders and dikes. Still, very large floods will probably one day again cause havoc down the valley.

Figure 27.4A–B. (*opposite*) Soda Dam and the old road in an undated photo, probably from the 1920s, showing a much wider Jemez River, possibly in near flood stage. The modern photo was taken in 2022. Other photo comparisons show that the riverbed in recent years is higher and narrower, with more willows and vegetation on the banks. (Early photo courtesy New Mexico History Museum, Palace of the Governors Photo Archives, NMHM/DCA Negative No. 057898; modern photo courtesy Miguel and Dee Plana)

Figure 28.1. Brachiopod and gastropod fossils from Madera limestone. (From Kues, "Guide to Late Pennsylvanian Paleontology")

CHAPTER 28

Fossils in the Jemez

The Jemez Valley is a great place to look for fossils in sedimentary rocks. When I was a young boy, I spent countless hours collecting brachiopod and crinoid fossils in the Madera limestone in our backyard. That rock layer was formed about 300 million years ago from coral reefs and an accumulation of animal shells in a shallow sea.[1] It is the gray to yellowish rock visible from the old mission church (Jemez Historic Site) up the canyon to Battleship Rock. Most of the stones in the walls of the old church are Madera limestone.

In chapter 2 I told fossil stories related to the Spanish Queen Mine. Over the past four centuries, miners dug out copper, silver, and gold ore from that spot in the red rocks located just south of Jemez Springs. The rock layer there is Abo sandstone and is about 290 million years old. In recent decades paleontologists have discovered many fossils of plants and animals within and around the old mine and farther down San Diego Canyon. The animal fossils include bones of a *Dimetrodon* species, which was a giant lizard-like creature with a tall, spiny sail on its back. It was not a dinosaur, and in recent years paleontologists have determined that mammals descended from a common line with *Dimetrodon*.[2]

The dinosaur era lasted from about 245 million to 66 million years ago. There are a few places with sedimentary rocks from that age in the lower Jemez Valley, but the best places for finding dinosaur bones are to the west of the Jemez Mountains. The Bisti Badlands/De-Na-Zin Wilderness area, for example, has extensive petrified wood from old forests, and many fabulous dinosaur fossils have been found there, including the bones and skull of *Bistahieversor sealeyi*. This predator was an older, more primitive relative of *Tyrannosaurus rex*.[3]

Although most residents and visitors to the valley have heard about New Mexico dinosaur fossils, fewer people know that more recent fossilized

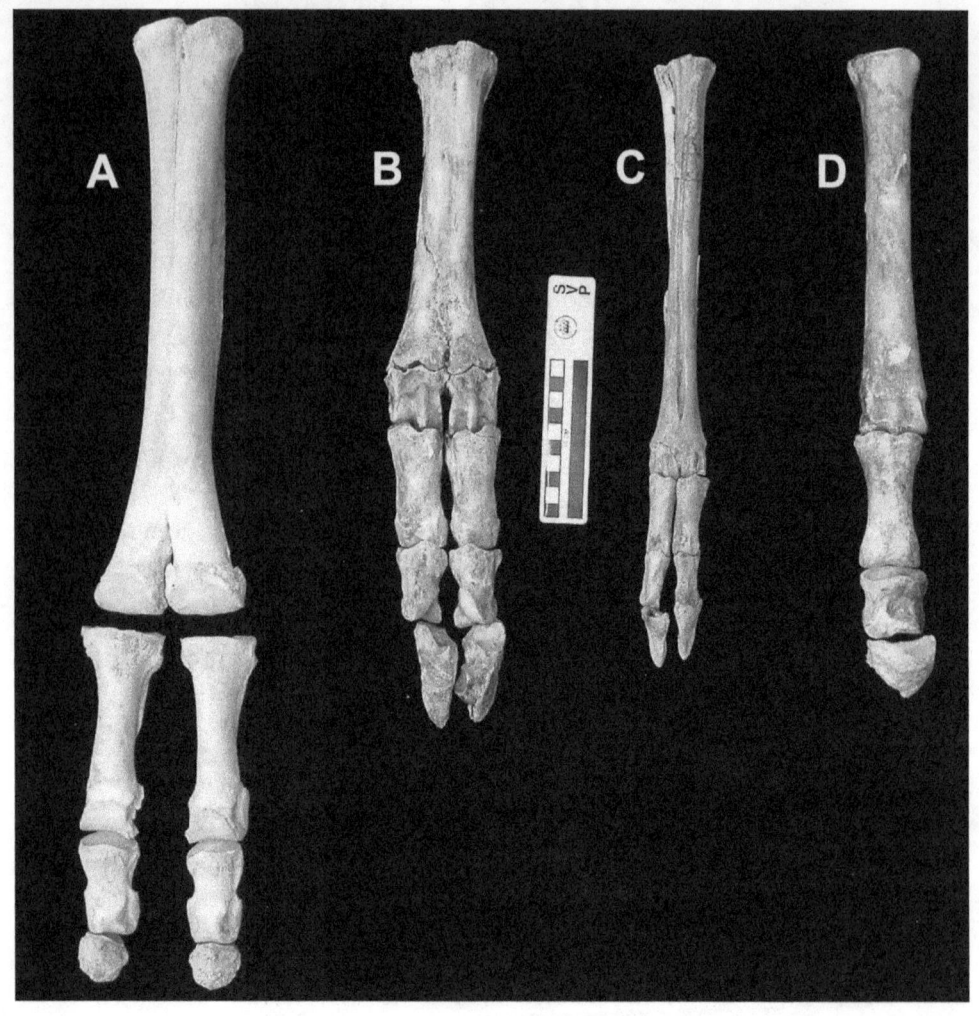

Figure 28.2. Leg and foot bones of camel, bison, deer, and horse from the White Mesa mine. (A) *Camelops hesternus*; (B) *Bison antiquus*; (C) *Odocoileus hemionus*; (D) *Haringtonhippus francisci*. (From Morgan and Rhinehart, "Late Pleistocene Mammals")

Chapter 28

animals from the Ice Age have also been found near and within the Jemez Mountains. The most recent Ice Age—the Pleistocene epoch—extended from about 2.58 million to 11,700 years ago. During the latter part of that epoch, many large and small animals that are now extinct roamed ancient grasslands near modern-day Jemez Pueblo, San Ysidro, and Zia Pueblo, and also the Chama River valley to the north of the Jemez, where mammoth bones and tusks were recently found.

These fossils, from the Hartley Mammoth Site just south of Abiquiu Reservoir, have been dated at about 37,500 years old using the carbon-14 dating method. This find is very interesting, and it's controversial too, because a group of scientists who studied the bones concluded that the mammoths were killed and butchered by humans. They claim that cut marks and other features of the bones suggest that humans processed the carcasses, and they found small micro-flakes at the site that they interpret may be from stone tools. These findings and interpretations of human involvement are not yet convincing to many paleontologists. In any case, it is clear that mammoths once roamed parts of the Jemez Mountains.[4]

Another set of remarkable Pleistocene-age fossils was found at the gypsum mine on White Mesa, located a few miles south of San Ysidro. This mine, operated on Zia Pueblo lands, uses heavy machinery to carve the bedded gypsum from the top of the mesa. The gypsum is processed and used in drywall boards. In 2005 mine workers noticed that their machine had dug up bits of old bones.

It turns out there were deep fissures in the gypsum deposit that served as natural traps. Large animals fell into the fissures, where they died and their bones were preserved. The bones were found about thirty to forty feet below the original surface of the gypsum deposit. Paleontologist Gary Morgan and several other paleontologists from the New Mexico Museum of Natural History and Science carefully excavated the fossils with help from Zia Pueblo mine workers David and Lambert Pino.[5]

Within three different fissures they found parts of the skeletons of a horse, a bison, two camels, and a mule deer. The bones were dated at 12,910 +/- 60 radiocarbon years old. (The calibrated calendar date is between 13,685 and 13,293 BCE.) The horse, bison, and camel species found are all extinct now. Only a related species of bison persisted in North America, while horses and camels dispersed in Asia and Europe. The species of horse found in the gypsum deposit was identified by Gary Morgan and Larry Rinehart as *Equus francisci*, the New World stilt-legged horse. Some years after their paper on the White Mesa fossils was published, other paleontologists demonstrated that this species is unique, and a new genus and species name was applied to it: *Haringtonhippus francisci*.[6]

Gary Morgan and Spencer Lucas published a lengthy summary paper on Pleistocene fossils in New Mexico in 2005. That paper includes remarkable mentions of Jemez fossils, including this one:[7]

> A single incomplete but intriguing fossil is known from Guadalupe Mesa on the west side of San Diego [Canyon], south of Jemez Springs in Sandoval County, NM. This fossil consists of a partial cervical vertebra of a large carnivore and is most similar to the comparable element in a large felid. The sabercat *Smilodon gracilis* and the cheetah-like cat *Miracinonyx inexpectatus* are two large felids typical of early Irvingtonian faunas and are of appropriate size and morphology to warrant further comparisons with the vertebra from Guadalupe Mesa. This fossil was collected from a pumice bed at the base of the upper Bandelier Tuff dated at 1.19–1.22 Ma.

The early Irvingtonian period was from about 1.6 million to 850,000 years ago. So to summarize, at different times in the distant past there were horses, camels, bison, mammoths, and probably sabertooth (or cheetah-like) cats in the Jemez.[8]

A recent paper published in the journal *Science* provides new perspectives on horses in North America. In this new study scientists and Indigenous tribal collaborators report the detailed investigation of horse bones in the Southwest and Great Plains. Carbon-14 dating indicates that at a Pueblo village site near Taos and in several states to the north and east domesticated horses were being used by Indigenous people by the early 1600s, if not earlier.[9]

This date is about one hundred years earlier than previous estimates of the major rise of horse cultures on the Great Plains. The old narrative was that extensive horse use there followed the 1680 Pueblo Revolt, when

Pueblo people sold and traded many horses seized from the Spanish to other tribes to the north and east. DNA testing of old horse bones from this time confirms that the horses descended from horses brought by the Spanish to the Southwest. However, native North American horse species persisted in Canada until about five thousand years ago and maybe later. So there is still a small possibility that bones of those horse species might eventually be found in association with ancient Indigenous people and sites.

Part IV

Fire, Forests, and Cottonwoods

A cottonwood in San Diego Canyon.

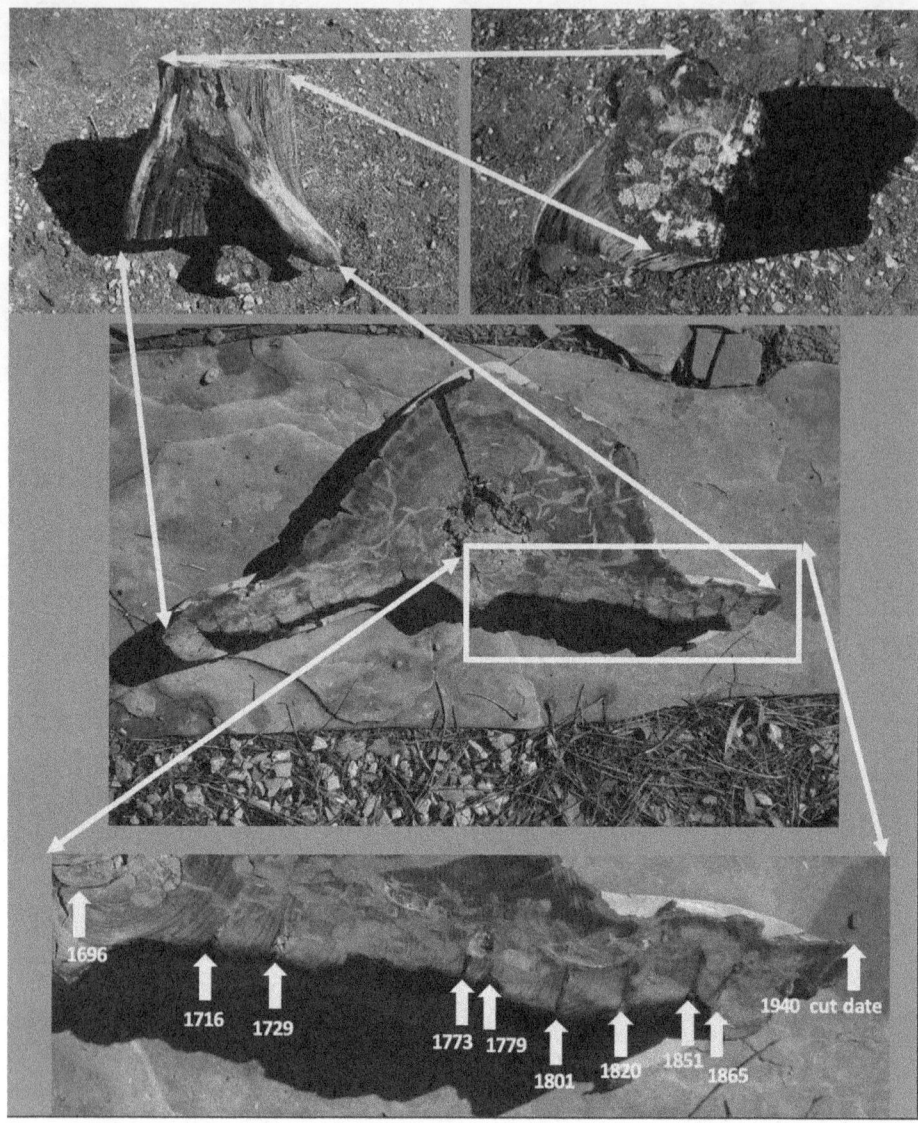

Figure 29.1. Cross section from the stump of a fire-scarred ponderosa pine showing nine fire dates from 1696 to 1865. Only the resinous heartwood portion of the stump remains. The tree was cut by crosscut saw and axe in 1940 on the bench above Area 3. Multiple fire scars are visible with arrows and dates. Beetle or wasp larval galleries were common in the Area 3 samples and created a challenge in crossdating ring patterns and determining which rings contained the fire scars. (From Swetnam, "Fire History of a Ponderosa Pine Stand")

CHAPTER 29

Forests and People on the Southern Jemez Plateau

Forest fires have always burned in the Jemez Mountains, and they always will. Even during years with very wet winters and plenty of snow, there are hot and dry days during the fire season, which usually extends from May through September. Pine needles, grasses, and weeds that grew abundantly in previous wet seasons dry out and provide fine fuel for any spark that will ignite a running wildfire.

Living in landscapes that are highly prone to burning every few years requires an ongoing rhythm of fuels management. All of us who wish to continue living in fire-prone woodlands and forests need to be fuel managers. We can learn some lessons about fuel and fire management from people who lived in large numbers in the canyons and on the mesas of the Jemez Mountains for centuries—the Hemish.

Starting around 2011, I was fortunate to work with a group of anthropologists, archaeologists, forest ecologists, and forest managers in learning about fire and human history in the Jemez. With funding from the National Science Foundation,[1] we collected tree-ring samples from dozens of places near and far from ancestral Jemez Pueblo villages on the mesas to the east, west, and north of Jemez Springs. We also carbon-14 dated charcoal in sediments to learn the history of fires over thousands of years. Finally, we interviewed Hemish elders and collaborated with Hemish cultural and natural resource managers to learn about their traditional knowledge of fire and forests.[2]

The southern Jemez Plateau, which includes the mesas and canyons south of Redondo Peak, is dotted with thousands of ruins of one- and two-room houses and dozens of midsize to large Hemish ancestral villages. Some of these villages had more than fifteen hundred rooms and were several stories

Figure 29.2A–B. Photographs of the changing fuel load at the Hemish ancestral village ruin of Amoxiumqua on Virgin Mesa. The first photo was taken in 1910, the second in 2016. Figures are standing in the same excavated room. (From Roos, "Fire Suppression Impacts")

tall. Recall what the big townhouses of Taos Pueblo look like today. From about 1400 to 1650, there were many townhouses like that on the mesas of the Jemez. Our best estimates from archaeological, documentary, and tree-ring studies is that in 1620, when Spanish missionaries and Hemish people built the Mission de San José, now the Jemez Historic Site, there were about five to eight thousand Hemish people living on the southern Jemez Plateau.[3]

For the most part, the pre-1600 villages were within ponderosa pine forests. These were, and still are, very fire-prone places, with grasses and pine needles normally accumulating enough fuel to support a spreading wildfire every few years. So, with lots of lightning and plenty of people setting fires, how did the Hemish people live within this fiery and smoky landscape for centuries?

Figure 29.3A–B. The first photo was taken in 1904 during a survey of the original Jemez Forest Reserve. The ponderosa pine forests on the south side of Redondo Peak, South Mountain, and Banco Bonito, and around Vallecitos de los Indios, had many more trees in 2023 (second photo) than in 1904, with fewer grassy openings, except where recent wildfires have burned intensely. The cleared areas near NM 4 in the left center were due to pumice mining (now ended) and the Cajete Fire of 2017. (1904 photo from Benedict and Reynolds, *Proposed Jemez Forest Reserve*)

Figure 29.4. Widely spaced, large-diameter ponderosa pines in a forest stand, probably located on Banco Bonito, 1904. (From Benedict and Reynolds, *Proposed Jemez Forest Reserve*)

Chapter 29

To shed light on this question, we used tree rings to learn when people built and lived in the villages and how frequently fires occurred near and far from the villages. The tree-ring evidence we gathered was of several kinds. First, using dendrochronology methods, we matched the ring-width patterns in roof timbers from the old dwellings with a master chronology of ring patterns from the Jemez Mountains. This allowed us to determine when the trees were cut and hence construction dates. Second, we took tree-ring samples from trees growing near and within the tumbled-down ruin walls and determined when they started growing there. Those dates gave us a minimum time since the village was depopulated. And last, we collected samples from old fire-scarred trees in many places, both near the ancestral villages and at increasing distances.

In a nutshell, here's what we found:[4] When the Hemish people lived in their villages before about 1650, they used thousands of small-diameter timbers for roof vigas in their townhouses, kivas, ramadas, and all sorts of other structures. They also gathered huge quantities of firewood over multiple centuries of living in these places. Almost all the largest villages are in winter-cold sites higher than seven thousand feet in elevation, so firewood gathering and burning were essential for cooking and heating. From determining tree ages immediately around and within the village ruins today, we found that very few trees were growing there during occupation as late as 1650.[5] Old photographs show that almost all the trees growing around the ruins today were established over the past 150 years, when livestock grazing and suppression of fires allowed forest ingrowth.

Furthermore, we found that at sites distant from the villages, there were many more small, low-intensity fires during the occupation period before 1650 than after that time. The high frequency of small fires suggests that in addition to lightning, people were setting fires, probably to clear their farm fields each spring, for hunting purposes, and to keep travel corridors open. This indicates that the Hemish people kept their villages safe from wildfire by harvesting small-diameter trees and gathering firewood, especially near their villages, and by setting many small fires to manage fuels in the surrounding landscape.[6]

And here's another important finding: Recent wildfires have been burning over ancestral pueblo ruins in the Jemez Mountains at unprecedented intensities, damaging potsherds and other archaeological materials remaining on the surface.[7] However, from all the archaeological, anthropological, and tree-ring evidence we have gathered so far, we have found no evidence that any of the Hemish ancestral villages were entirely burned or abandoned after burning in wildfires.

The Hemish evidently lived sustainably within forests for centuries. Contrast that accomplishment with what has happened here in the Jemez Mountains and elsewhere in the western United States in recent decades, with hundreds of homes and entire towns going up in wildfire flames. The increase in extremely severe wildfires in dry pine-dominant forests in recent decades is due to the accumulation of living and dead woody fuels over the past 150 years of fire suppression, warming temperatures, and drought. In published studies of the role of climate in driving recent wildfires in the western United States, scientists have concluded that the warming and drying of fuels due to rising greenhouse gases in the atmosphere is responsible for about half of the increase in area burned.[8]

So how can we reduce wildfire risk to our homes in woodlands and forests? Following the "firewise" lead of the Hemish would be an excellent start. Clear the fuels around your home and neighborhoods. Also, support the thinning of small-diameter trees, especially around and within neighborhoods and in critical watersheds that contain the primary springs, pipelines, and water tanks of our domestic water supplies. Prescribed burning is also essential, including broadcast fire and pile burning. It is not practical or affordable to treat all areas with only chainsaws and hauling off the wood, so using thinning in combination with low-intensity fire is an effective alternative. This also means that, like the Hemish, we must learn to live with both fire and smoky conditions during parts of some years. Of course, fire must be used very carefully in the right place and time. We can do this. The Hemish did, and forest managers are successfully employing these strategies in forests of the Jemez Mountains and across the western United States.

CHAPTER 30

The Era of Runaway Wildfires

On June 5, 1971, I was pumping gas at the Chevron station and grocery store across from Los Ojos Bar when I looked up and saw a mushroom cloud rising over Virgin Mesa. At first I thought maybe Los Alamos had been nuked! I quickly realized the smoke wasn't coming from that direction. In fact, the smoke plume was from a forest fire burning somewhere on the mesa southwest of La Cueva. Over the next hours and days, the Cebollita Fire burned more than thirty-five hundred acres. At that time this was considered a large wildfire. It burned so hot and fast the first day that it seemed it would torch all the forest and cabins from La Cueva to Coyote.[1]

In hindsight, I realize that this fire began the era of runaway wildfires in the Jemez. Before 1971, going back to about 1910, there were very few fires larger than one thousand acres in the Jemez Mountains. In the more than fifty years since the Cebollita Fire, we have had many wildfires larger than four thousand acres; the 2011 Las Conchas Fire was 156,000 acres. There are two primary causes of this sharp upward trend: accumulated fuels—living and dead trees—during the fire suppression era of 1910 to 1970 and increasingly hot drought conditions in recent decades related to the rise of greenhouse gases in the atmosphere.[2]

As mentioned in previous chapters, my father, Fred Swetnam, was the Jemez District ranger in 1971. After hearing the fire report from Cerro Pelado Lookout, he rushed in his green Forest Service truck up to Fenton Hill. As soon as he saw the giant flames, he radioed the Santa Fe National Forest dispatcher and requested aerial retardant drops. A World War II–era high-altitude B-17 bomber, converted to a low-altitude slurry bomber, arrived within about thirty minutes. However, the first drop had little effect, and the turnaround time for the next drop was more than an hour.

By the next day the fire was more than two thousand acres in size. More than one thousand firefighters were on the scene battling the blaze,

Figure 30.1A–B. The Cebollita Fire approached a cabin in La Cueva around June 6, 1971. (Courtesy Swetnam family)

along with five slurry bombers, several helicopters, and multiple bulldozers. Then the winds died down, and the fire was contained on the sixth day. No homes were burned, and no one was seriously injured.[3]

The fire might have been stopped earlier because local loggers had arrived quickly with bulldozers and begun cutting lines around the fire. But when the Forest Service's Project Fire team (now called Incident Command) arrived and took over leadership from the district, they pulled most of the local dozers off the line because they did not have the required safety features, which included a steel cage around the operator. One operator had already been knocked off his dozer by a falling tree, but he wasn't seriously injured. Still, there were hard feelings afterward by some who thought the Forest Service had slowed the response.[4]

On June 17 my dad was quoted in the *Albuquerque Tribune*:

"From first glance, I knew it was a bad one." Ranger Swetnam says that at its height, this great Jemez fire reached such fierce heat that it created a "fire storm." "It was like a monstrous serpent writhing and twisting in the air," he said. "Imagine several giant whirl winds of fire whirling at great speed and also moving forward. The roar was like a big train. Flames would race along the ground and suddenly be sucked up to the top of a tall tree." Mr. Hall [Santa Fe National Forest supervisor] believes that he saw flames that were 1,000 feet high.[5]

The Cebollita Fire was also featured in the July 1971 issue of *Life* magazine, with a photo of my dad walking through a burned-out Jemez forest and a photo of Ron Brown, a longtime Jemez resident who was a young firefighter at the time. Again, my dad was quoted in his evocative manner of speaking about this fire: "The wild beast is out there. We know it. The Indian gods up in the hills must be laughing at us for trying to contest it."[6]

The whirlwinds my dad described are characteristic of wind-driven fires and the most extreme wildfire behaviors. Giant vortices, or "firenados," have become relatively common in today's megafires.[7] On the first afternoon of the 2011 Las Conchas Fire, a pair of immense counter-rotating whirlwinds developed along the flaming front, with the tilted-over smoke column rising more than thirty thousand feet in height. That fire burned about forty-three thousand acres on the first day, more than ten times the size of the 1971 Cebollita Fire, which burned for more than a week.[8]

The aftermath of these increasingly large, very hot wildfires is a depressing sight to behold. Where once there was a continuous canopy of green forest, now there are thousands of burned tree trunks, many still standing, and blackened ground, burned down to mineral soil. In following seasons and years, these burned places do come back to life and greenery.

Fire is as natural as wind and rain, and forests are generally well adapted to recovering from most fires. However, today, fires in the Jemez and elsewhere in the West tend to burn too hot and in too large patches. Trees can reestablish if there are surviving mother trees nearby to provide seeds. But some of the canopy holes created by recent wildfires in the Jemez are more than ten thousand acres in size.[9]

Ponderosa pine seeds are relatively heavy and don't disperse far from a mother tree. After these hot fires kill nearby seed trees, grasses and shrubs dominate quickly, outcompeting tree seedlings for light and moisture. The hotter and drier conditions of recent decades have also caused the death of

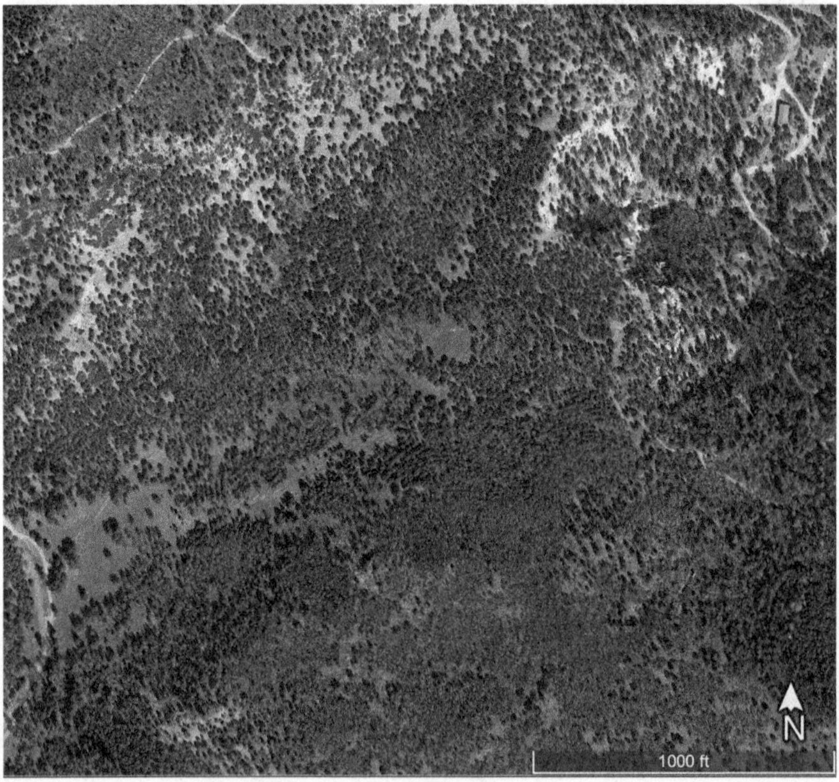

most tree seedlings that have been established. Our central problem now has become what forest ecologists call "type conversion," the permanent conversion of forests to shrub fields and grasslands.[10]

The Jemez Ranger District foresters of the 1970s were determined to help the burned-over areas of the Cebollita Fire recover back to forests. They carried out an aggressive agro-forestry approach, which included contour-plowing parts of the burn with tractors and planting thousands of tree seedlings in moisture-retaining terraces. We first learned about this fifty-year-old effort when we studied the aerial lidar (laser scanned) remote sensing images of the Jemez, which clearly show the plowed contour lines on the ground. Likewise, a Google Earth view of this area just south of La Cueva shows the now grown-up trees in neat rows. Someone told me later that they called these Jemez reforestation efforts "Fred's Farms."

The fifty-year-old Cebollita Fire plantations now need a thinning because the planted trees are growing too close to one another. However, this experiment seems to be a forgotten relic of twentieth-century reforestation; there are no signs of forestry thinning in the stands now, and no thinning is planned for the future to my knowledge. That is a pity because we need to learn how to help reforest and then manage recovered forests in the Jemez over the long term, including learning from past reforestation attempts, even if they were too heavy-handed. There is a forest there now, and if it is to survive the next fire, it will need some careful thinning soon.

Most foresters and the public today would agree, I think, that reforestation efforts at broad scales should not transform burned-out forests into plantations, with trees planted in regimented rows of tractor-plowed contours. Still, as the climate continues to warm, critical watershed protection and the survival of animal and plant species dependent upon forests may call for novel or even radical approaches to reestablish canopy cover. For example, planting more heat- and drought-tolerant tree species not native to these mountains might be an experimental option if native species reforestation efforts fail.[11]

For the immediate future, it is clear that some kind of reforestation with native species, guided by the best available science, needs to be undertaken

Figure 30.2A–B. (*opposite*) A lidar image of the ground surface shows plowed contours on the mesa immediately south of La Cueva. The same view from Google Earth shows a plantation of ponderosa pine trees growing in the terraces created by the plowing.

Figure 30.3. These fifty-year-old planted ponderosa pine trees, growing along plowed terraces on the mesa just south of La Cueva, were part of a reforestation project following the 1971 Cebollita Fire.

in large burned-over areas. Fortunately, the New Mexico State Legislature and the governor recently passed and signed a spending bill that provides $8.5 million to begin developing the New Mexico Reforestation Center and another $1.5 million for revegetation needs in the state.[12] That's an okay start, but much more investment is needed considering that the cost of recent wildfires in New Mexico is now in the multibillion-dollar range.

CHAPTER 31

Bald Mountains in the Jemez

Norteños know that the common adjective *pelado* means "bald" or "bare." Jemez Mountains residents recall Cerro Pelado as a grass-topped mountain with an old lookout tower that the Forest Service has used for more than a century to spot wildfires on the south and east sides of the range.

Almost all the highest peaks that circle and rise within the Valles Caldera are pelado on their south sides. Montane grasslands grow there from about ninety-five hundred to eleven thousand feet in elevation. They are (or were) ancient, treeless slopes with unique bunch grasses that in wet years grow thick and tall, up to belly height of a big horse.

The Hemish have long recognized an eagle image on the south side of Redondo Peak. The bird's head, beak, and wings are formed by the grasslands there.[1] At least they were until wildfires in recent decades killed many trees edging the iconic grassland image.

When the US Army's Wheeler Survey mapped the "Sierra de los Valles" (the Jemez Mountains) in 1876, they used the name Pelado for modern-day Redondo Peak.[2] Sometime later the name was changed to Redondo Peak in government maps, and the name Cerro Pelado was applied to the lower peak to the southeast. In Towa, the Hemish language, the name for Redondo Peak is Wâavêmâ. The translation of that name is "Want for Nothing Peak," and it has deep cultural and historical meaning for the Hemish people.[3]

In 2022 the Cerro Pelado Fire burned the lower slopes of this bald peak. The old lookout tower was saved, and thanks to previous fuels treatments, so were about two hundred homes to the north and west in the Sierra de los Piños neighborhood. The winds pushed the fire east and south into the 2011 Las Conchas Fire scar and then north toward Los Alamos.

In 2000, eleven years before the Las Conchas Fire (see the previous chapter), the Cerro Grande Fire burned into Los Alamos after it was initially

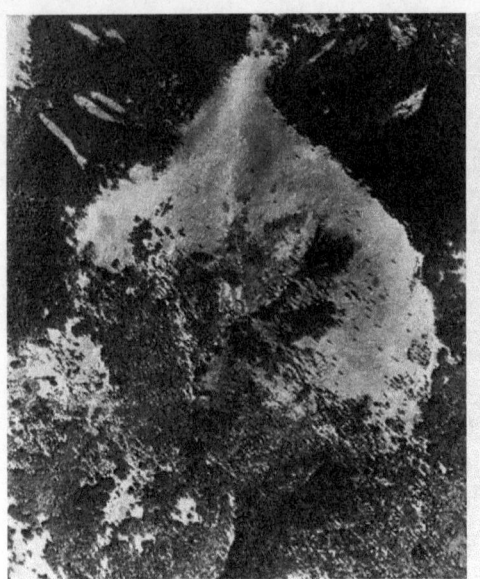 

Figure 31.1A–B. These comparison photos show the Cerro Grande grassland beginning to be invaded by trees in 1935 (*left*) and much more invasion by the late 1970s (*right*). (From Allen, "Montane Grasslands")

ignited as a prescribed fire just north of NM 4. Most people in Los Alamos were mandatorily evacuated during both events. For a while in 2022 it seemed that Los Alamos would be evacuated again, on an eleven-year cycle. But fortunately, the winds decreased and the fire slowed down. Eventually, the Cerro Pelado Fire was declared contained and then extinguished.

For over a year the cause of the Cerro Pelado Fire remained a mystery. The Forest Service said an investigation was under way and wouldn't comment further. Then, in late July 2023, it released the report. Two separate investigative teams had concluded that the fire was "possibly" ignited by holdover heat and sparks in large slash piles from a thinning project on San Juan Mesa. The Forest Service had carried out the pile burning in January and February and may have failed to detect the smoldering fires after declaring the pile burns out.[4]

The finger-pointing toward the Forest Service's failures was intense. This was particularly so because at about the same time the Cerro Pelado fire ignited, two other fires ignited by the agency—a broadcast burn and another holdover pile burn—ignited two fires on the east side of the Sangre

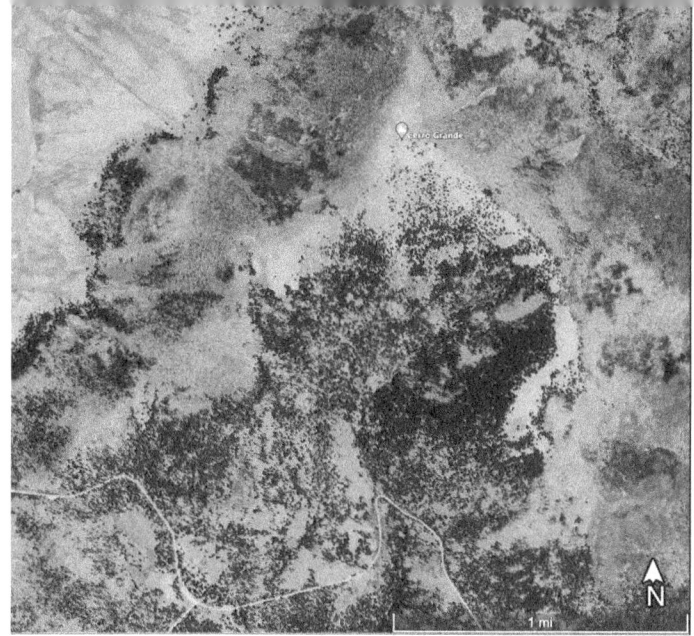

Figure 31.2. This image from Google Earth, taken in 2017, shows the Cerro Grande grassland further invaded, while much of the surrounding landscape has burned and converted from forest to shrublands. NM 4 is visible at the lower left.

de Cristo Mountains near Las Vegas, New Mexico. These fires merged, becoming the largest wildfire in New Mexico state history—the Hermit Peak-Calf Canyon Fire.[5]

So it would seem the Forest Service was to blame. However true this was for the original lighting of these fires, the full blame extends far and wide, encompassing decisions by people and government agencies for more than a century. I'll come back to the "blame game" later. For now, I continue here with further ecological background on bald Jemez mountains and fire.

My close friend and colleague Craig Allen has studied the Jemez Mountains for more than forty years as a scientist with the National Park Service and the US Geological Survey. He looked closely at the Cerro Grande grasslands and on other Jemez Peaks for his master's thesis work in the 1980s.[6] He discovered that the grasslands on the south aspects of the high peaks are thousands of years old.

Beneath them are grassland-type soils several feet deep that could only have been formed over many centuries. The grassy balds were maintained by frequent surface fires, burning about once per decade until around 1900. After intensive sheep and cattle grazing in the early 1900s and the end of frequent low-intensity fires, these pelado slopes began to be invaded by trees.

Here is where the story becomes layered with irony. In 1999 Craig and I, along with another longtime colleague, Julio Betancourt, published a scientific paper that proposed that managers could use historical evidence of past landscapes to recognize undesirable changes that had occurred. The loss of the Cerro Grande grassland was an example we used. Aerial photos from 1935 compared with those from the late 1970s from Craig's thesis showed how the open grassland and savanna had partly transformed into a closed-canopy forest over five decades.[7]

Restoration of the ancient pelado grassland on Cerro Grande was one of the justifications the National Park Service used for prescribed fires there. Another was reduction of fire hazard to Los Alamos. Fire managers repeatedly tried to burn this slope with planned burns in the 1980s and 1990s. However, the intensities of those fires were not high enough to kill many invading trees, as intended.

In 2000 the National Park Service tried a prescribed burn in early May, and the rest is history. There is much more to this story, and there are multiple interpretations of how it all happened. The shortest version is that extreme winds came up, the fire escaped, and it burned part of Los Alamos. It cost the US government (taxpayers) about one billion dollars.[8]

The causes of and responsibility for the "wicked problem" of accumulated forest fuels, climate change, and homes built within overgrown forests that burn too hot are broad and collective. There is plenty of blame to go around.

A great irony is that today, the former pelado slopes of Cerro Grande and Cerro Pelado have more forest than ever. In fact, there is more forest canopy cover there than on most of the surrounding mountainsides, which were repeatedly burned over with high intensity by the 2000 Cerro Grande Fire, the 2011 Las Conchas Fire, and the 2022 Cerro Pelado Fire.

We now know that the Cerro Pelado Fire might have been caused by a holdover, a reburn from smoldering slash fuels ignited by the Forest Service months before. The Forest Service has proximate responsibility for this disaster today, but the causes and responsibilities go way back and extend beyond a government agency. They include decisions by private landowners

Figure 31.3. A view from the grassland on Redondo Peak looking northeast toward the Valle Grande. Visible in the foreground at lower left is Thurber fescue, a very robust native bunch grass that occurs on *pelado* southern exposures of Redondo and other Jemez peaks.

to run too many livestock on this landscape for decades, which inhibited the spread of beneficial low-intensity fires. The blame also goes to government policies of putting out all wildfires for more than one hundred years. The blame also goes to loggers, who cut most of the big trees but then failed to remove all the slash fuels or to continue good forestry by thinning the dense reproduction of small-diameter trees that followed the logging.

So what have we learned? What can we learn? History is essential to understanding how we arrived at our current situation, but as unprecedented change becomes the norm, the past becomes less useful as a guide to the future.

Still, given our combined problems of abundant flammable fuels, many homes built within our forests, and a warming and drying climate, our rational options do not include doing nothing. The science and the experiences of forest and fire managers across the country have shown that forest thinning and prescribed burning are effective and safe tools when used in the right places and times.[9]

Figure 32.1. View from our deck in Area 3 toward the east and Cat Mesa. The sundial arrow shows the direction of the spread of the Cerro Pelado Fire at about 5 p.m. on April 22, 2022.

CHAPTER 32

Cerro Pelado Fire and Using All the Tools

We have two choices for the kinds of wildfires that will burn in the Jemez Mountains: those that are destructive to homes, forests, and watersheds or those that are beneficial, sustaining our landscapes and lives. We do not have the option of no fires. We tried putting out all fires for more than a century, and this just made matters worse. Too many trees grew in our forests, and too much dead fuel accumulated on the forest floor. These fuels burn too hot and fast when wildfires occur.

Forest thinning with machines is essential to remove some of the fuels, but we can't solve the fire problem entirely by cutting our way out of it. Now, with rising temperatures and hundreds of homes built in the woods, we are facing the consequences of not using all the tools available to us, including both chainsaws and fire.

We saw again the potential for wildfire disaster on April 22, 2022, when high winds pushed a very fast, intense wildfire straight toward Sierra de los Pinos, a development with about two hundred homes in the forest south of the Valles Caldera National Preserve. When I first saw this fire from my deck that afternoon, I thought those homes would be burned up within hours. Instead, the thick plumes of black smoke from the raging fire slowly turned mostly white and gray. The fire had lost some energy, shifting toward the east and northeast, away from the community. Amazingly, only a few structures were lost on the first day with epic winds.

Why didn't the fire blaze through the middle of Sierra de los Pinos and burn up dozens of homes? It wasn't because of immediate firefighting actions. Because of the extreme winds and dryness, the wildfire blew up too fast to catch it small. Firefighters and emergency responders had their hands full helping people evacuate, and the winds were too high for aerial

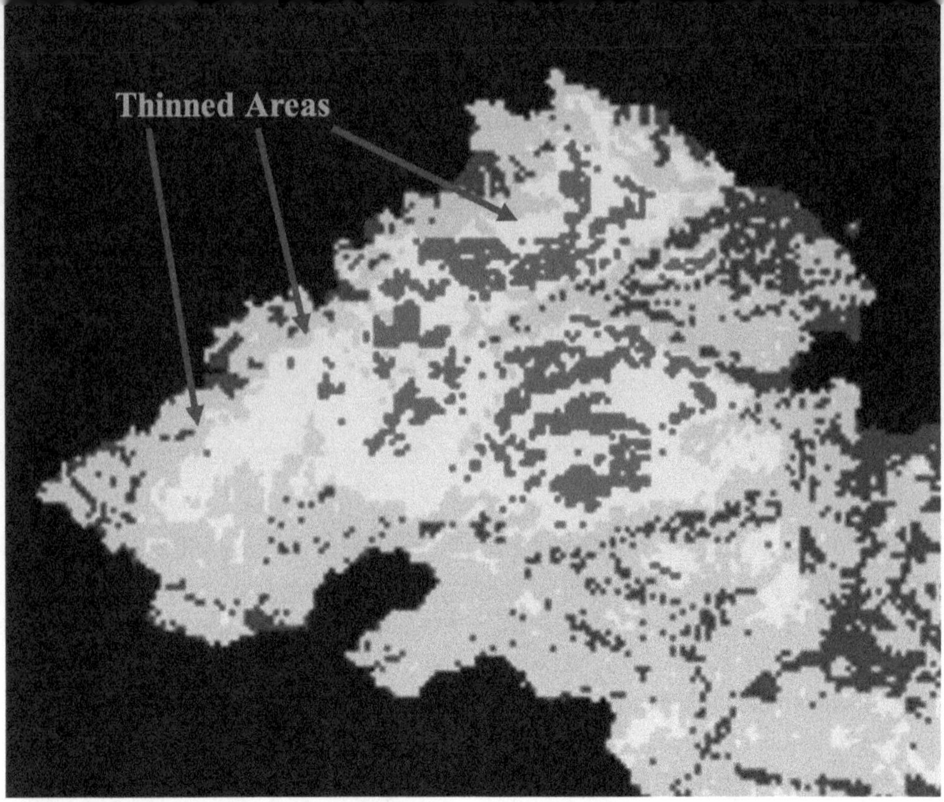
Thinned Areas

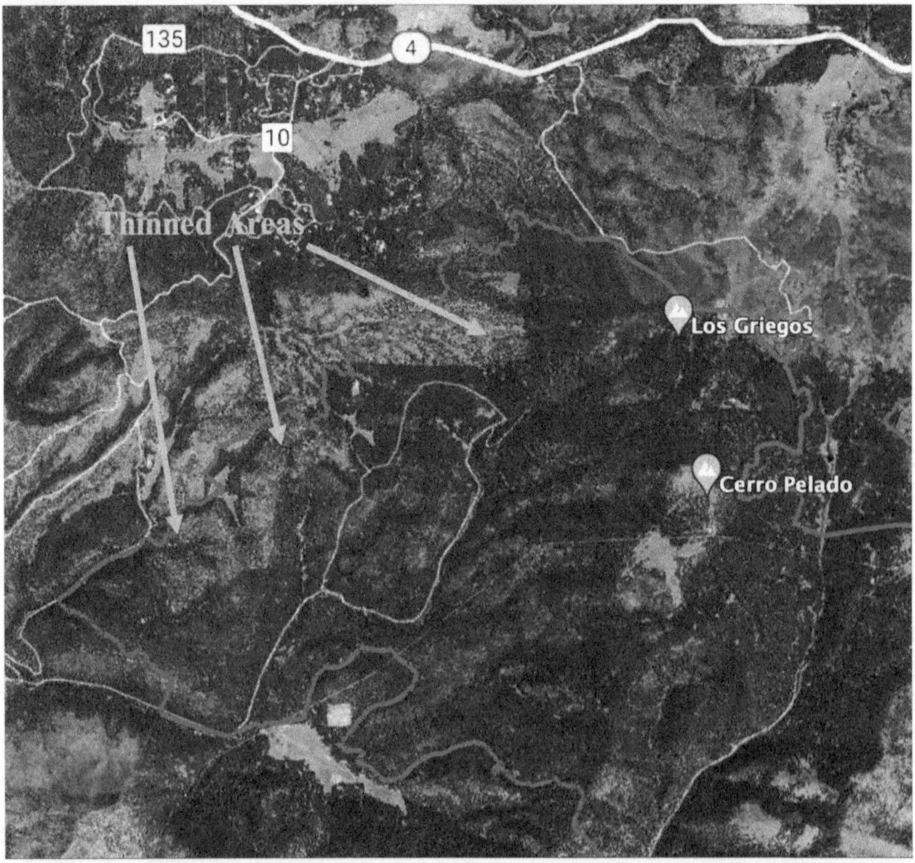
135
4
10
Thinned Areas
Los Griegos
Cerro Pelado

Chapter 32

retardant planes and helicopters to fly. The answer is that extensive forest thinning and fuel treatments on the mesas and slopes south of Sierra de los Pinos probably made the difference.

The running crown fire that was pushed from the south by wind gusts moving more than thirty miles per hour likely dropped out of the treetops to the ground when it hit the broad swath of thinned forest. The winds also cooperated by shifting the fire more to the east. The less intense fire front then skirted east and northeast, mainly burning around the neighborhood.

The Cerro Pelado Fire is not the only example where forest thinning and fuel treatments helped save many homes and possibly human lives. In 2018 the Venado Fire, burning northwest of Guadalupe Box, was a very hot running crown fire burning northward toward the Chaparral Girl Scout Camp. The fire raged down into a canyon and up a south-facing slope. Then it hit the ridge top, where an intensive thinning and pile-burning project had just been completed. The crown fire dropped to the ground and became a surface fire that was much easier to stop. If the thinning and pile burning had not been done, the Venado Fire might have rolled all the way to Cuba.

Unfortunately, we now know from multiple examples that escaped prescribed fires and holdover pile burns can become very destructive wildfires. In the summer of 2023, we learned that the Cerro Pelado Fire was another case where months of residual burning in slash piles might have provided the sparks for igniting this wildfire in the very dry and windy April 2022.

My colleagues Matt Hurteau, Craig Allen, and I wrote the following op-ed while multiple escaped fires were burning in New Mexico in April 2022. Despite the errors of escaped fires, we argued then, and still hold now, that we must not abandon using fire to fight fire: we need all the tools available to us.[1]

> Figure 32.2A–B. (*opposite*) The first image shows the burn severity in the northwestern corner of the Cerro Pelado Fire, where it started, on the first day. Color coding: medium gray = high severity; white = moderate severity; light gray = low severity; dark gray = unburned. The second image is a 2020 Google Earth view of this landscape. The thinned areas generally burned at lower severities than the unthinned areas. (First image from Burned Area Emergency Response program, US Geological Survey, https://burnseverity.cr.usgs.gov/baer)

Figure 32.3. Part of the thinned area south of Sierra de Los Pinos, showing effects of the Cerro Pelado Fire. The fire dropped to the ground in most places and scorched the lower canopies of trees but did not kill them. In other places, patches of trees in the thinned area were killed. Post-fire erosion has occurred in some places, as the gully in the foreground shows. The photo was taken in September 2022.

Fire can cause wonder or fear, renewal, or destruction—and, at this moment, there is considerable fear in some of our communities.

While current wildfires started from various sources, their impacts on our communities and ecosystems were determined well before the ignitions. The unnaturally large amounts of fuel in our forests resulted from societal choices over the past 100 years as we actively worked to exclude even natural beneficial fires from our forests. The current megadrought has been made worse by our widespread and

Chapter 32

prolonged burning of fossil fuels, and the resulting warming climate has conditioned forest fuels so they are more flammable. Just add an ignition and an extreme wind event, and impactful wildfire can ensue.

We provide this larger context because it is central to recognize that we are all in this together because of our collective choices over the past century. As we, here in New Mexico and across the Western United States, work to reduce the wildland fire problems we have created and the risks they pose to our communities, we have to accept that our problem-solving tools, mechanical thinning and prescribed fire, are not without their own sets of challenges and risks.

A century-long policy of putting out all fires helped cause our widespread forest and fire problems in the first place, and for many reasons, we can't fix the fuels problems solely with chainsaws. Beneficial fire is the most cost-effective, practical, and ecologically appropriate way for restoring many of our forests.

Escaped prescribed fire like the Hermits Peak Fire calls for careful investigation and reevaluation of decision processes, and adjustments as needed. But arm-chair quarterbacking and clamoring for accountability before a proper after-action review will not improve prescribed fire use. Instead, it will only serve to reinforce the already existing risk aversion of land managers, leading to inaction and further increases in our risk from wildfire.

Prescribed fire escape is extremely rare, and abandoning its use as a tool to restore forest resiliency because of an escaped fire would be akin to abandoning medical interventions because sometimes doctors make mistakes.

We face a wildfire problem fueled by fire suppression and climate change. The fire season is lengthening as winter shortens. Fires are burning hotter and moving faster as our winter snowpack decreases. Fire suppression crews used to make considerable progress on containing fires at night because they slowed and became less active due to lower temperature and higher relative humidity; this often is no longer the case.

Some wildfires are burning as actively at night as they do during the day. Even seasoned wildland fire professionals and research scientists, like us, have been surprised by how quickly wildfire behavior has changed. We are all in this together because we live, recreate and are sustained by these fire-prone forest watersheds.

Maintaining New Mexico's forests requires restoring fire, which will help make our forests resilient to climate change. We need to acknowledge and accept that fire managers working to restore our forests and decrease wildfire risk are doing so on our behalf within many constraints to address a substantial problem that we have collectively created.

When a prescribed fire escapes, learning is essential—we should be patient and let a thorough review determine any avoidable errors or needed changes to our decision systems. We need to carefully choose the times and places to use fire, as well as know when and where not to burn.

In any case, it is clear we need all available tools to restore forest resilience and protect communities in the face of climate change, including the judicious use of fire.

CHAPTER 33

Cottonwoods and Junipers in the Valley

A few years ago, a Forest Service crew of sawyers removed a couple stands of juniper trees growing along the valley bottom of San Diego Canyon. Someone raised concerns about that action in the local newspaper, pointing out that junipers and cottonwood trees are both native to this landscape but that junipers use less water. If the cottonwood bosque expands farther because competition with junipers is decreased, water flows may decrease further in the Jemez River as thirsty cottonwoods thrive. Water for irrigation and maintaining the river ecosystem is critical, especially in the extraordinarily hot drought we have experienced in recent decades.[1]

Cottonwood bosques also become highly flammable as trees age and as leaves, dead branches, and stems accumulate. For example, very hot fires break out nearly every year in the Rio Grande bosque. We are likely to have such fires in the Jemez Valley, sooner or later, and expansion of cottonwoods and junipers along the river corridor may increase the chances for wildfires.

When considering forest and woodland management, it is usually insightful to study the long-term history of a landscape. Trees live a long time and are greatly affected by natural and human-caused disturbances. Fortunately, we have several ways to reconstruct the history of the Jemez River bosque. We can determine tree ages using tree-ring studies. We can examine the size and distribution of trees using lidar and photographs, including aerial photos.

A set of spectacular photographs from 1884 shows the Jemez River and corridor.[2] These photos show that cottonwoods were almost absent from long reaches of the Jemez River, where they now occur as dense and tall stands of trees. Most striking in these photos is the relatively barren condition of the river bottom in 1884, with very few trees of any species

Figure 33.1A–B. Views of the junction of the Rio Guadalupe and Rio Jemez at Cañon. The first photograph is from a series taken in San Diego Canyon in 1884. The modern photo was taken in 2015. (1884 photo courtesy Library of Congress LC-USZ6–975)

Figure 33.2A–B. View looking north up San Diego Canyon, about midway between Cañon and Jemez Springs, 1884. If you look closely, you can barely see a two-track wagon "road" winding along the east side of the river. The modern photo was taken in 2015. (1884 photo courtesy Library of Congress LC-USZ6–973)

Figure 33.3A–B. Looking south from a high point directly above the center of the village of Jemez Springs at the bathhouse. The 2015 photo was taken from the lower bench, just visible at the bottom center of the earlier photo (1884 photo courtesy Library of Congress LC-USZ6-974).

growing on the floodplain. The questions arise: Why were no cottonwoods and few junipers growing along these parts of the valley, and why did the cottonwoods subsequently establish in large numbers?

More study is needed to know the precise reasons and chronology of events, but several factors were probably at play at different times and magnitudes. Flooding is certainly a major player in the ecological dynamics of cottonwood bosques. Large floods can rip out trees, but the seeds and roots of uprooted cottonwood trees are then spread across new sand and gravel bars, where they sprout prolifically. This is a primary way that cottonwood bosques establish.[3]

The great flood of 1958 was responsible for one of the most common cottonwood cohorts (tree groups of the same age) growing along the Jemez River corridor today. You can see this cohort especially clearly in the meanders of the flooding river that washed out NM 4. These erosion features are evident along the east side of the road in several places below Jemez Springs, and they are filled with cottonwood trees that established after the 1958 flood or possibly the 1941 flood.

However, there were also multiple large floods in the late 1800s, as evidenced by old photos of Soda Dam showing massive log jams emplaced around the dome over the Jemez River (see chapter 27). Despite those floods and earlier torrents, cottonwoods were absent from long stretches of the Jemez River in 1884.

One possible reason for the absence of extensive cottonwood cohorts of any age in the 1884 photos is livestock grazing. Vast numbers of sheep were grazed in the Jemez Mountains from the late 1870s to the 1920s.[4] Large bands of sheep were moved up and down the Jemez Valley seasonally. Sheep are especially fond of eating newly sprouted cottonwood seedlings. So it is likely that heavy and repeated grazing pressures kept cottonwoods from successfully sprouting and resprouting along the Jemez River following floods.

One other possible reason for the absence of cottonwoods, and for that matter the lower density of junipers or piñons in the 1884 pictures, is fuelwood gathering. Very large quantities of fuelwood were needed in the nineteenth, eighteenth, and earlier centuries, when this was the only source of fuel for heating and cooking. It was not uncommon for people to travel many miles to gather wagon- and donkey loads of fuelwood because, after centuries

Figure 33.4A–B. Rio Guadalupe just below the box. The 1944 photo shows the effects of the 1941 flood, which tore out SFNW railroad tracks. The second photo was taken in 2018. (1944 photo courtesy US Forest Service, Jemez Ranger District; 2018 photo courtesy Steve Sutherland)

of use, woodlands were completely depleted near the villages. Cottonwood is not the most prized fuelwood, but it ignites easily and burns hot.

Ultimately, repeat photography and other historical information demonstrate that our Jemez landscape is highly changeable. Droughts, floods, fires, tree cutting, and livestock grazing have resulted in waves of change, sometimes resulting in loss of forests and sometimes their rebirth. In this context the term *restoration* is highly ambiguous. Restoration to what condition? To what time?

I think the answer, in this landscape that has for so many centuries been both a natural and a cultural landscape, is the condition that promotes natural and cultural values. The first priority in watershed management along the Jemez River corridor should be removal of non-native trees that have been introduced. Another priority is to reduce water loss to vegetation in the watershed because our communities depend upon irrigation water. Most notably, Russian olive trees and tamarisks are present in many places, and these should be eradicated, if possible, as they are non-native and thirsty trees.

In the past century junipers and cottonwoods have expanded in places in the Jemez Valley. If we want more water and less fire risk, a trade-off will be less exotic *and* natural vegetation (fuels) along the river. In this sense, the recent juniper cutting makes sense; these areas are not likely to convert to cottonwood until another big flood occurs.

One last comment, and a mea culpa: I am partly responsible for the presence of Russian olive trees along the Jemez River south of Jemez Springs. As a young Jemez Springs Boy Scout in the 1960s, I recall planting Russian olive seedlings as a service project. My recollection is fuzzy, but I think it was a project supported by the USDA Soil Conservation Service. The purpose was soil erosion prevention and wildlife habitat enhancement. It goes to show a truism of land management history: The road to hell is paved with good intentions.

This history also evokes Aldo Leopold's dictum: "A thing is right when it tends to preserve the integrity, stability, and beauty of the biotic community. It is wrong when it tends otherwise."[5] I would add that the biotic community includes rivers and people.

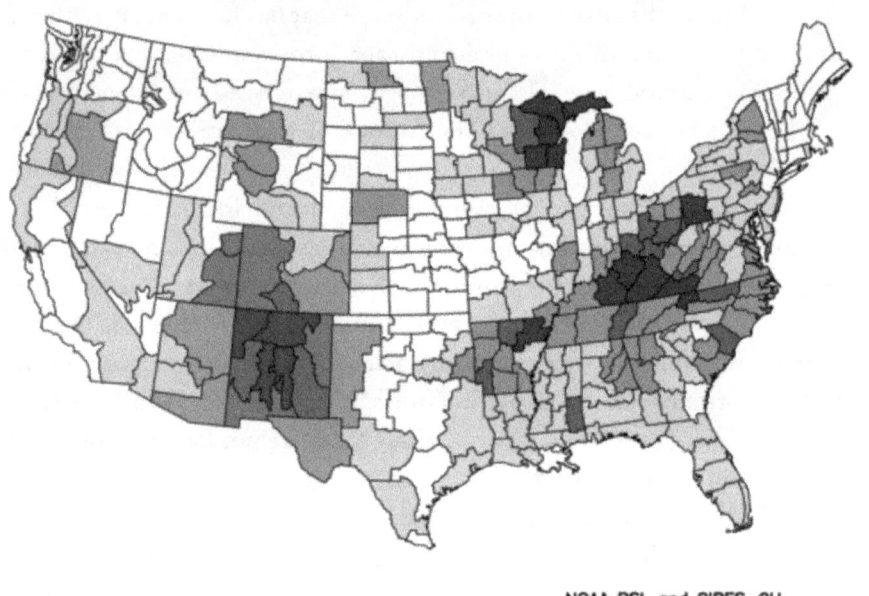

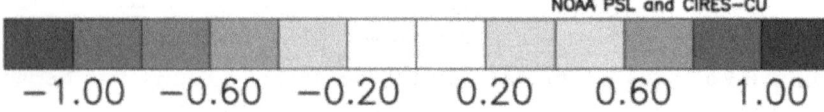

Figure 34.1. Rainfall in the Southwest is much reduced, especially in the Rio Grande Valley in New Mexico from 2015 to 2023 compared to preceding decades. On the map, shaded areas west of the Mississippi are below average (drier, negative values), and shaded areas east of the Mississippi are above average (wetter, positive values). (From National Oceanic and Atmospheric Administration, National Center for Environmental Information)

CHAPTER 34

Climate Change in the Jemez

In previous chapters I mention the reality of climate change and that the resulting hot drought of recent decades is one of the factors driving huge forest fires in the Jemez. When I talk about our fire and climate history work in the Jemez in public meetings, I bring up the fact that the recent drought, since about 2000, is extraordinary. I point out that highly credible scientists have concluded that human-caused greenhouse warming of the atmosphere and the consequent drying of fuels are responsible for nearly half of the increase in area burned by wildfires in the western United States.[1]

I don't think my statements about what scientists have concluded are very convincing to some. But almost everyone agrees that it has indeed been hotter and drier in recent years. This is clear to most people because they have experienced the hot drought of the past decade-plus. So I begin the discussion of climate change by showing facts—that is, actual measurements of temperature and precipitation here in the Southwest and the Jemez Mountains.

The map of precipitation anomalies shows that the Southwest has been ground zero for drought since at least 2015. Time series from the Jemez Mountains reveal the same pattern, especially for spring through summer temperatures and precipitation. These data indicate that the hot drought here began as early as the late 1990s.

In 2012 my colleague Park Williams led the analysis and writing of a scientific paper (I was a coauthor) on the hot drought of recent decades in the Southwest and the strong correlation of drought indices with increasing forest areas burned by wildfires and killed by drought and bark beetles. In this paper Park developed a new metric of drought, specifically focusing on the effects of heating and drying on forest growth and tree death. We call this the Forest Drought Stress Index (FDSI).[2]

FDSI is computed using a combination of cold-season precipitation (positive relationship) and warm-season vapor pressure deficit (VPD, negative

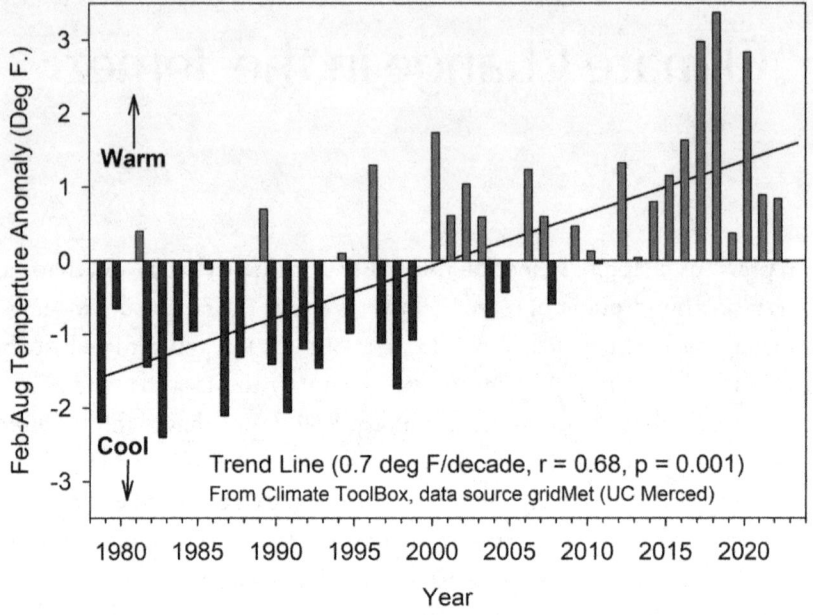

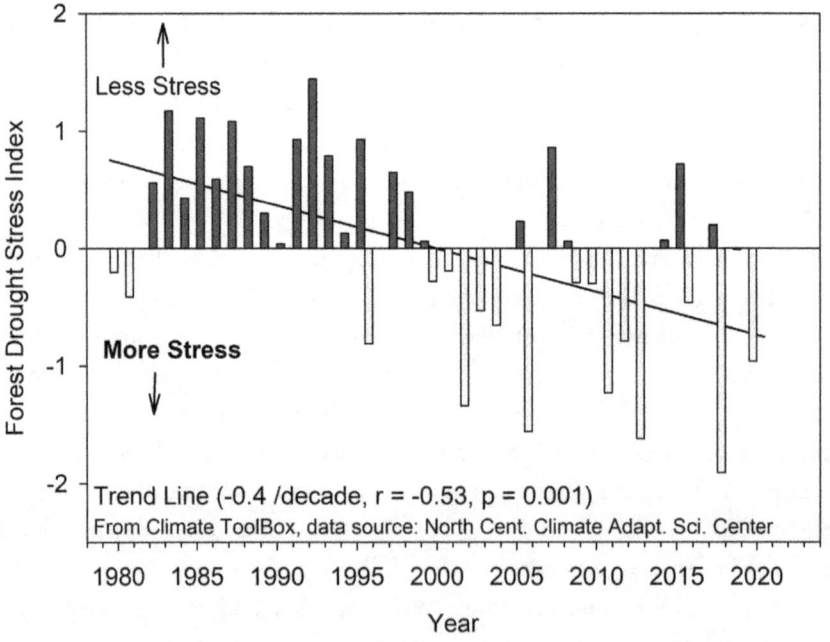

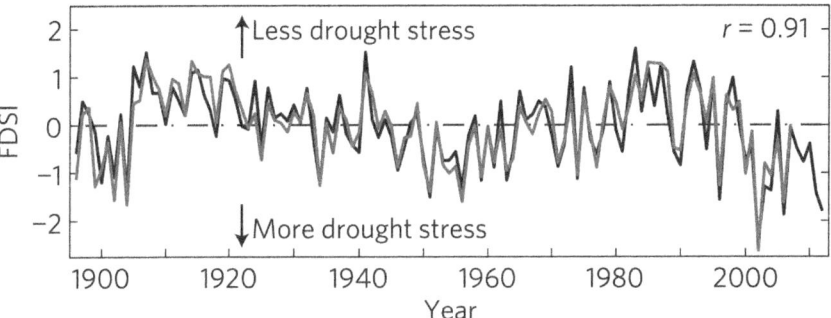

Figure 34.2A–B. (*opposite*) Temperature and precipitation data from a rectangular area enclosing the Jemez Mountains, Nacimiento Range, and San Pedro Parks Wilderness. The data, used to compute the Forest Drought Stress Index (FDSI), are from weather stations within the Jemez River watershed. The values plotted are anomalies—the differences between annual values and the average of values from 1979 to 2023 for temperature and from 1980 to 2020 for FDSI. (From Climate Toolbox, Applied Climate Science Lab, University of California, Merced)

Figure 34.3. (*above*) The annual FDSI for Arizona and New Mexico computed from instrument measurements of temperature and vapor pressure deficit (black line, 1896–2012) and the annual FDSI estimated from tree-ring width records (gray line, 1896–2007). (From Williams et al., "Temperature as a Potent Driver")

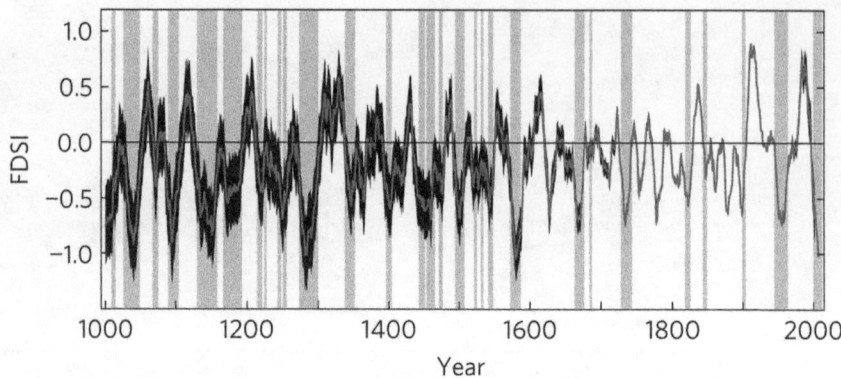

Figure 34.4. FDSI time series (*gray line*) from 1000 to 2007 CE, estimated from 335 tree-ring width chronologies from the southwestern United States. The black areas are 95 percent confidence intervals representing the range of FDSI values expected if all 335 chronologies were available. (A declining number of tree-ring chronologies were included in the dataset in earlier periods.) The vertical gray bars show the major droughts of the past millennium. Notice the 1950s and 2000 droughts compared to the 1580s drought. (From Williams et al., "Temperature as a Potent Driver")

relationship). VPD measures the "dryness" or "thirstiness" of the air. VPD reflects how much more moisture the air could hold before becoming saturated, essentially indicating the difference between the actual water vapor pressure and the saturation water vapor pressure at a specific temperature. The VPD used to compute FDSI is a combination of the previous August to October and growing season, May to July, VPD.[3]

We found that annual FDSI was very well correlated with tree-ring width growth from hundreds of forest sites sampled across the Southwest and with observations of annual forest areas killed by disturbances. Basically, with a combination of rising temperatures, rising VPD, and precipitation decreases, FDSI decreases as well (more drought stress), trees have reduced ring-width growth, and more trees die in larger areas from drought, bark beetles, and wildfires. Figure 34.4 shows a perspective of FDSI changes over the past one thousand years estimated from tree-ring width chronologies.

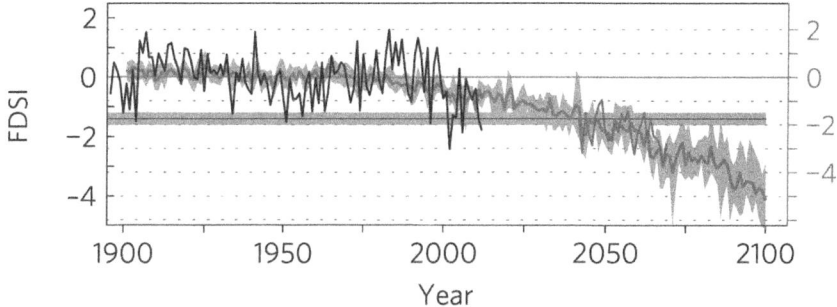

Figure 34.5. Projected FDSI using outputs from an ensemble of global circulation models, as driven by external forcing, including business-as-usual scenarios of increasing greenhouse gases in the atmosphere through the twenty-first century. Color coding: *black*: observed records; *gray line*: CMIP3 ensemble mean values; *shading around gray line*: inner 50% of CMIP3 values; *dark gray time series*: 2042–2069 dynamically downscaled NARCCAP ensemble mean values; *horizontal gray line and shading*: mean and 95 percent confidence FDSI values of the most severe 50 percent of years during the 1572–1587 megadrought. Horizontal gray dotted lines show the anomaly in standard deviations from the observed 1896–2007 mean (*right y-axis*). (From Williams et al., "Temperature as a Potent Driver")

The one-thousand-year perspective of FDSI from tree rings shows that major droughts and wet periods have recurred repeatedly over past centuries. The most extreme droughts in the past lasted more than a decade and had great impacts on people and ecosystems. For example, great droughts in the mid-1100s and the late 1200s contributed to the displacement and migration of Puebloan people within the region. The 1580s megadrought also led to movement of people from the Pajarito Plateau in the Jemez to wetter places along the Rio Grande.[4] And there was probably also extensive die-off of trees during that megadrought.[5]

When we published this graph and paper in 2012, the 2000s drought was almost as extreme as the 1580s megadrought. Since that time the drought has continued, and recently Williams and colleagues showed that the current southwestern drought is the most severe drought in the past twelve hundred

Chapter 34

years. In general, it is evident that the recent drought in the Southwest is largely driven by warming temperatures.[6]

The final analysis described in the 2012 Williams et al. paper is a look at the potential effect of a continued rise of greenhouse gases in the atmosphere (from fossil fuel burning) on FDSI through the twenty-first century. This analysis used the outputs of a set of global circulation models (running on supercomputers) that predict temperature and precipitation changes down to the subregional scale. The model predictions, in this case, were based on business-as-usual scenarios of continued greenhouse gas emissions.

The results of the modeled prediction of FDSI over the coming century are sobering. Sometime in the 2030s to 2050s, the *average* climate conditions are predicted to be hotter and drier than during the 1580s megadrought. Given the increasing magnitude and extent of forest die-off and the lack of regeneration of forests in many places, this climate change trajectory is not compatible with continuation of forests as we know them today in the Jemez. The impacts of such changes on water resources and people will also undoubtedly be very severe.

When I describe this rather grim outlook that data, models, and climate knowledge are telling us in local public talks, I can usually sense that some people are not convinced that climate change is caused by humans, specifically that greenhouse gases are responsible for changes we have seen recently. In distrusting what most credible scientists have concluded, some also dismiss the computer model predictions of continued warming and drying in this century. A common reaction I hear in conversation with those who doubt that the projections are reliable is, "Well climate is cyclical, and this is just a dry [or warm] period, and sometime in the future the climate will turn wet [or cooler] again." This is wishful thinking.

It is true that climate goes up and down—that is, it oscillates between warm and cold, wet and dry. Some of the most important oscillations, such as the El Niño-Southern Oscillation (ENSO), have become relatively well

Figure 34.6A–B. (*opposite*) View of Cerro Picacho on the Pajarito Plateau, with the Sandia Mountains near Albuquerque in the distance (*at upper left*). The first photo was taken prior to the 1996 Dome Fire, which burned over Cerro Picacho. In 2011 the Cerro Grande Fire reburned this landscape. Now only a few small patches of trees remain. (Courtesy Craig Allen)

understood. The global ENSO pattern affects the southwestern United States with typically wetter cool seasons (winter to spring) during El Niño events and drier ones during La Niña events. Periods between oscillations vary from about two to seven years. The wet-dry switching on this time scale results in a moderate correlation of fire occurrence (area burned) in the Southwest with ENSO. But the patterns are not predictable more than a year or so in advance, and the relationship with wildfire has weakened in recent years.[7]

Note that the term *oscillation* is used rather than the term *cycles*, which implies that the time periods between the peaks and valleys of the changes are the same, and likewise the magnitudes. However, the times between ENSO events of El Niño and La Niña, and their strengths, are too variable to be considered cycles. Likewise, there are other ocean-atmosphere oscillations that affect climate in the Southwest on time scales of decades to multiple decades, but they are generally even less predictable and have less effect on our climate than ENSO.[8]

In any case, the general warming trend due to greenhouse gases is now surely affecting these climate oscillation patterns. The increasing temperature at a base level has begun to override the increased cool-season moisture we tend to get during El Niño events. That is because the thirstiness of the atmosphere (as measured in VPD, for example) increases exponentially with increasing temperature. As the atmosphere warms, there is an accelerating rate of moisture evaporation from soils and vegetation.[9]

The effect of global warming of the atmosphere and oceans on ENSO dynamics is not well understood. In some ways, ENSO dynamics and mechanisms are more complicated than greenhouse gas effects on warming, and the latter is now affecting the former. Even if we were to have more frequent and stronger El Niños in the coming decades, which might mean more cool-season precipitation in the Southwest, the increasing rate of evaporation due to continued warming would reduce the net amount of moisture received and retained in trees and watersheds. We might also have less or more frequent La Niña events, but the increased warming will exacerbate the drying effects that usually attend La Niñas in the Southwest.

Although this is bad news, most scientists I respect say there is still time to reduce the trajectory of warming and prevent the worst effects on ecosystems and people. But time is getting short, and we need to act more quickly than we have so far to reduce greenhouse gas emissions.[10] The modeling results shown by Williams et al. are computed using business-as-usual scenarios—based on assumptions that we will not quickly transition away from fossil fuels to renewable energy sources in the next few decades and that we won't invest extensively in carbon-capturing strategies, such as reforestation and protecting remaining old growth forests.

A good friend of mine who was a talented basketball player in high school and college told me that our attitude toward beating the climate change challenge must be one of "confidence in winning." Winning athletes know that a combination of effort, talent, and confidence is essential. I think it is also important to understand and face up to the challenges.

Part V

Hotels, Hot Springs, Hippies, Campers, and Priests

Battleship Rock.

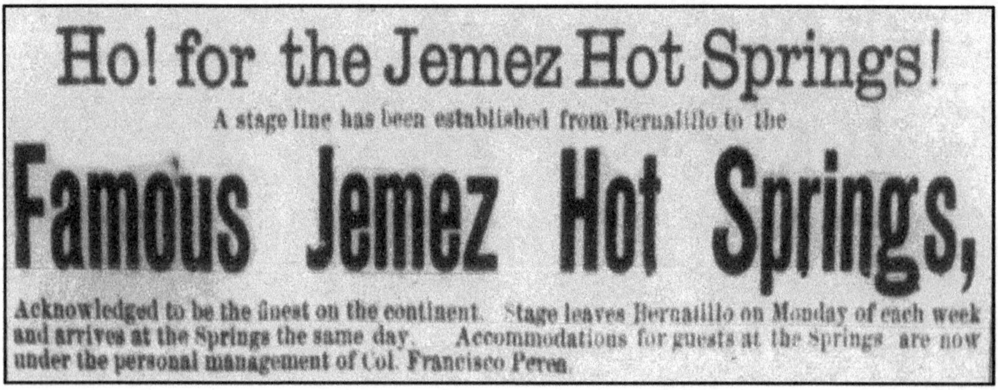

> **Ho! for the Jemez Hot Springs!**
> A stage line has been established from Bernalillo to the
> **Famous Jemez Hot Springs,**
> Acknowledged to be the finest on the continent. Stage leaves Bernalillo on Monday of each week and arrives at the Springs the same day. Accommodations for guests at the Springs are now under the personal management of Col. Francisco Perea.

Figure 35.1. Albuquerque newspaper advertisement in 1881 for the stage and accommodations (at the Stone Hotel), with the bold claim, "Acknowledged to be the finest on the continent."

CHAPTER 35

Hotels, Mines, and The Sulphurs

After the railroad arrived in Las Vegas, New Mexico, in 1879 a tourism boom followed. The AT&SF began building hotels at the most attractive destinations along the rail lines. Among the most desirable destinations were hot springs with adjacent hotels that offered private rooms, running water, and good food. The hot mineral water baths were promoted for their medicinal and salutary effects on everything from eczema to tuberculosis. In 1881 the AT&SF built the large Montezuma Hotel just north of Las Vegas at a hot springs site, and it promptly burned down. The railroad rebuilt it, and it burned again in 1885, but reflecting the investor's confidence in its ultimate profitability, it was promptly rebuilt a second time.[1]

The ambitious businessmen Mariano S. Otero and his uncle Miguel A. Otero thought Jemez Hot Springs was also destined to be a major tourist attraction. When word got around that the AT&SF was considering building a rail spur to Jemez Hot Springs, the two Oteros jumped at the opportunity and purchased several large land tracts in the village, including the main hot springs. Colonel Francisco Perea, Mariano's cousin-in-law, joined them as a third investor. They planned to build a hotel and bathhouse and cash in on the flood of tourists coming by rail to the "Yellowstone of the Southwest," as the Jemez was hyped in newspapers. The Oteros and Pereas were heirs to a small portion of the Baca Location No. 1 land grant, and by the late 1890s, Mariano had bought up about 30 percent of the grant from the other heirs.[2]

This short announcement of the Jemez Hot Springs purchase appeared in the *Las Vegas Gazette* on December 11, 1880:[4]

> It is announced that the honorable M.A. Otero has purchased a third interest in the Jemez Hot Springs 30 miles from Bernalillo for $15,000. These springs are rated as among the best medicinal springs in the world and an analysis recently made shows them to be as good if not

better than our own springs. The other owners are Delegate M.S. Otero and Colonel Francisco Perea of Bernalillo. A large hotel with all the modern improvements will be erected and also a large bathhouse. Money will be freely expanded to make it an attractive resort, and printers ink lavishly used to bring it into favorable notice.

The Stone Hotel, with about a dozen private rooms and all the amenities, was constructed in 1881 along with the bathhouse. Today the old hotel is a large, stone-walled building on the Bodhi Manda (Zen Buddhist) property. After the hotel and bathhouse were built, the famous Colonel Perea—a veteran of the Battle of Glorieta Pass, a territorial delegate to Washington, DC, and friend of Abraham Lincoln—moved to Jemez Hot Springs with his large family and managed the operations. As it turned out, the rail line was never built to Jemez Springs, but regular stagecoach runs from Bernalillo delivered a smaller number of tourists over the decades.[5]

Mariano Otero's ambitions were not limited to hotels, bathhouses, and sheep grazing on the Baca. In the late 1890s he purchased a mineral claim on the western boundary of the Baca with sulphur deposits visible on the surface. These deposits on the west side of the Sierra de los Valles were known to early Spanish colonizers, who called them Los Azufres (The Sulphurs).[6] Numerous fumaroles in the area emitting hydrogen sulfide gas for centuries had deposited crystalized, nearly pure sulphur at ground level and below.

To mine and haul the valuable sulphur, Otero poured money into improving the old "military road" that came up the Pajarito Plateau from the east to the caldera rim, approximately where NM 4 is today. He built dozens more miles across the valles and down into the canyon on the west side where The Sulphurs were located. In 1902 teamsters using ten wagons brought the first load of mining equipment up the road from Santa Fe. The largest freight wagon had two teams of eight horses each, front and back, pulling and pushing a heavy boiler for the steam engine that would run the ore processing plant. From 1902 to 1904 more than two hundred thousand pounds of sulphur were processed and hauled out of the Jemez.[7]

In addition to the mine works, Otero built a nearby hotel and a bathhouse that enclosed some hot springs. For many years afterward, tourists were transported by stage line over the mountain roads from Thornton (six miles south of Peña Blanca) up to the gold mining town of Bland and then

Figure 35.2A–B. The Sulphur Springs area in a photo from 1905, with the old hotel on the far left in the shadows of trees. A modern photo shows the area in 2022. (1905 photo courtesy New Mexico State University Library Archive, Special Collections, Image No: 01040347; 2022 photo courtesy Miguel and Dee Plana)

over the caldera rim to the Sulphur Springs resort. That must have been a very long, bumpy, and scenic ride! Another stagecoach line ran from Bernalillo through Jemez Hot Springs and then up the canyon to La Cueva and finally to The Sulphurs.

Otero's dream of a rich return from the sulphur mine was short-lived. After the initial diggings of sulphur crystals near the surface played out, the miners found they needed mine shafts to get at deeper deposits. That's where they ran into trouble from heat and poisonous hydrogen sulfide gas. The newspaper article below, from May 22, 1903, quotes Mariano Otero describing their challenges. He also reports a most amazing discovery of ancient Spanish diggings on the site.[8]

LIKE WORKING IN A FURNACE

> Terrific Heat in Jemez Sulphur Mine Resembles Miniature Hades

OLD SHAFT DISCOVERED

> Dug by the Dons Centuries Ago—Boiler Explosion Hurts Three Men—Dead Loads of Sulphur

M.S. Otero came in yesterday from Jemez Springs. He has been up at his sulphur mines near that resort for some time superintending operations which are going forward steadily and rapidly.

"We made a peculiar find the other day while at work on our new shaft," said Mr. Otero. "After excavating for 10 feet or more we struck the mouth of an old shaft which is without doubt a relic of the old Spaniards. It was covered over with rotten timbers and for the first twenty feet the hole was clear. It was neatly timbered up and braced, the wood however being so rotten that it crumbled away in the hands. There is no record to show that any American or native has ever had a mine in that locality and it is evident that the shaft is one made by the Spaniards centuries ago and covered up at the time of the Pueblo revolution, as were many other mines throughout New Mexico. The condition of the shaft shows that it was a relic of antiquity. What the Spaniards were after there is a mystery as they probably had no use for the sulphur. [But note: sulphur was a primary ingredient for gunpowder.]

This sulphur mine is going to be one of the famous mineral properties of the Southwest. The trouble with development work now," said Mr. Otero, "is that when the men get down 75 or 100 feet it gets so intolerably hot that they cannot endure it. The sulphur is responsible for this extremely high temperature and makes labor in the mine like work in a furnace. We were unable to explore the old shaft any deeper just on this account as the men were nearly suffocated after staying in the hole a short time. But when more new shafts and air tunnels are opened up however, we expect to be able to get much deeper."

Several days ago a boiler at the sulphur plant exploded with terrific force. Luckily the men were all at a safe distance, although three of them were painfully injured. On account of the boiler and necessary repairs to machinery they are not doing much this week but expect to be hammering away in a few days.

The mines are just a few miles this side of the springs where there is a regular mountain of sulphur. Hundreds of pounds of pure fine sulphur have been taken out and when everything is ready, Mr. Otero expects to be able to take tons of the odiferous stuff out every day. The supply seems inexhaustible and from all appearances the Jemez sulphur mine is going to be a famous one. The company is already employing a large force of men and has an outfit of new and modern machinery with which to handle the output. Sulphur is valuable and that the mine is going to be a mint goes without saying.

By 1905 it was apparent that the easily accessible sulphur was exhausted, and the heat and poisonous gas problem in the deep shafts brought the mining to an abrupt halt. The hotel and bathhouse operated for decades after that. The property changed hands multiple times, and by the 1970s a sort of hippie commune existed on the site. The old hotel burned down around that time. A deep geothermal exploration well was drilled on the site in the 1970s.[9]

> # Sulphur Springs Stage Route
> ### Via BLAND in the GOLDEN COCHITI,
>
> The best equipped four-horse stage line in the Southwest, from Thornton to the famous Sulphur Springs in the Jemez mountains.
>
> ### THE SCENIC STAGE ROUTE OF NEW MEXICO.
>
> Leave Thornton Tuesdays, Thursdays and Saturdays at 8 a. m.; arrive at Bland at 12 m. Leave Bland at 1 p. m. and arrive at Sulphurs at 6 p. m. Stage returns from Sulphurs on Mondays, Wednesdays and Fridays.
>
> The new management of the Sulphur Springs ha provided
> ### FIRST CLASS HOTEL ACCOMMODATIONS
> and reconstructed the Baths and employs competent attendants.
>
> Tickets for sale at W. L. Trimble & Co.'s stable at Albuquerque, and agent at Thornton.

Figure 35.3. Newspaper advertisement in the *Santa Fe New Mexican*, August 5, 1898.

A longtime resident of the valley who worked for the exploration company at the time recalled an incident at a nearby well (Redondo Border) when workmen were knocked unconscious by hydrogen sulfide gas, but fortunately they were revived.[10] In January 2020 the National Park Service acquired the property with help from the US Land and Water Conservation Fund and several private donors. Managers at Valles Caldera National Preserve are considering restoring some aspects of the site for public viewing. Hot springs, boiling mud pots, and fumaroles might be visited along boardwalks, for example, such as at Lassen Volcanic and Yellowstone National Parks.[11]

I wonder if any remnants of the old Spanish *azufre* mine might be found someday.

CHAPTER 36

The Many Camps at Battleship Rock

One of the most iconic landmarks in the Jemez Mountains is Battleship Rock. The Jemez River begins here at the confluence of the Rio San Antonio and the East Fork of the Jemez River. Battleship itself is a narrow convergence of two cliff faces, appearing like the prow of a warship. It was formed by the filling in of a paleo-canyon with volcanic rock (ignimbrite) that flowed out of the El Cajete vent at the base of Redondo Peak more than sixty thousand years ago (see chapter 26). The two rivers then excavated the sediments on the sides of the narrow volcanic cast, leaving what is considered by geologists "a beautiful example of reversed topography."[1] The volcanic flows from El Cajete vent, South Mountain, and San Antonio Mountain dammed up both the Rio San Antonio and East Fork on multiple occasions, creating temporary lakes in the great valles upstream (see chapter 23).

Adjacent to the confluence and right beneath Battleship Rock are relatively flat sedimentary benches that have been favorite living and camping sites for hundreds if not thousands of years. Several Hemish ancestral villages are within easy walking distance, and dozens of rock shelters and one- to four-room "field house" ruins are scattered around the flat areas and nearby slopes.[2] The best-known camp at this site today—and most beloved by generations of young campers—is the YMCA's Camp Shaver.

The written history of the campsite begins in the 1880s, when James Smith homesteaded at Battleship Rock and constructed a water-powered sawmill there. Smith was an early Anglo-American settler in the Jemez beginning sometime after 1876. For a while he lived at Sulphur Springs, where he probably had a cabin and bathhouse for guests.[3] In 1882 he married Calletana Archuleta Williams, a granddaughter of José and Maria Archuleta, who were among the earliest San Diego Land Grant settlers at Jemez Hot Springs in the early 1850s.

Figure 36.1. University of New Mexico Battleship Rock Camp in 1930. (Courtesy Maxwell Museum of Anthropology, University of New Mexico, Catalog No. 2009.24.8)

James and Calletana had many children, and their descendants and relatives include families still living in the Jemez Valley today. From the early 1880s until the early 1900s, Smith carried out logging on the slopes and benches in San Diego Canyon, using horses and oxen to drag the logs to chutes and landings, and then to the sawmill. (See chapter 21 on 1880s logging in Area 3.)

In a 1979 interview James's son Clyde Smith related that land surveys were carried out in about 1912 after the San Diego Land Grant was partitioned and sold. It was determined that the sawmill was located on the forest reserve side of the boundary, not on the San Diego Land Grant. Consequently, Smith had to discontinue his sawmill operation there.[4] The national forest-land grant boundary line passes through the middle of where Camp Shaver is located today. There is no clear sign of the old sawmill buildings except for some low stone and concrete walls close to the river.

After about 1912 the abandoned sawmill site was probably a favorite campsite of fishermen, hunters, government surveyors, and forest rangers. The first mention in New Mexico newspapers that I can find about using this place

Figure 36.2. View of the UNM archaeology field camp taken from the top of Battleship Rock in 1930 or 1931. Note the canvas tents pitched around the clearing and several buildings on the west end (*right center*). (Courtesy Maxwell Museum of Anthropology, University of New Mexico, Catalog No. 2009.24.11)

as a large group campsite is dated August 23, 1926.[5] This news story tells of 150 Boy Scouts making the adventurous automobile trip up the sandy, rutted dirt roads from Albuquerque to Battleship Rock. They called this place Camp Porter.

Perhaps Mr. Porter was a supporter of the Boy Scouts at that time. The newspaper articles don't clarify the origin of the name, but I suspect it is associated with Guy Porter, who was one of the White Pine Lumber Company owners and railroad logging managers on the San Diego Land Grant in the Rio Guadalupe drainage. A logging camp at the junction of the Rio de las Vacas and Rio Cebolla at that time was also called Porter or Porter Landing (see chapter 12).

The Boy Scouts returned seasonally to Camp Porter at Battleship Rock during the late 1920s to the early 1930s. Camping at that time by Boy Scouts and others was primarily in canvas tents arrayed around the margins of the large clearing in the forest where the old sawmill had been. During these

Figure 36.3. La Junta Fly Camp buildings of the Civilian Conservation Corps at Battleship Rock, 1936. (Courtesy Southwestern Region, US Forest Service)

years the Boy Scouts erected several buildings, which were probably used as a kitchen, mess hall, and headquarters. All those buildings are gone now.[6]

During this same period (at least as early as 1930), the University of New Mexico began holding summer archaeological field schools at what it called the Battleship Rock Camp. The university probably had some kind of cooperative relationship with the Boy Scouts for coordinating use of the camping areas and buildings. Professors and students from UNM and many other universities around the country camped in tents while excavating the nearby Hemish ancestral village ruins of Unshagi and Nanishagi. These Pueblo ruins are located north and south, respectively, of what is today Hummingbird Music Camp (established in 1959).[7]

In 1933 the Civilian Conservation Corps (CCC) was established as part of the New Deal, and hundreds of young men were employed on national forest and national park lands. They built roads, campgrounds, lookouts, and stone and wood structures, and did erosion control and forestry projects. The first camp in the Jemez Mountains was probably located at La Cueva. Other camps in the Jemez were built later during the program, including "fly camps" at Valle Grande (on the upper East Fork at Las Conchas), Paliza (on Vallecitos Creek), and Battleship Rock (site of La Junta Fly Camp).[8]

The La Cueva CCC camp enrollees were involved in building parts of NM 126, creating in combination with NM 4 what was called the Trans-Jemez Road. The Las Conchas and Paliza CCC enrollees built Forest Road 10 and other roads providing new access through the Vallecitos drainages for hauling logs and lumber from timber cut on the Baca Location, the national forest, and the San Diego Land Grant. In 1936 CCC workers at La Junta Fly Camp excavated and restored parts of the seventeenth-century Mission de San José ruins in Jemez Springs (see chapter 2). Around this time they also improved Battleship Campground and built the octagonal stone pavilion that is still there today.

Another New Deal program, the National Youth Administration (NYA), arrived in the Jemez in 1938. This program was championed by Eleanor Roosevelt and was intended to complement CCC employment. This was also a work program, but in contrast with the CCC, the NYA was targeted for younger (eighteen to twenty-five years old) unemployed men from families receiving some kind of public assistance. Employment by NYA was seasonal in the Jemez.

Apparently, the CCC's La Junta Fly Camp at Battleship was disbanded or moved before 1938, and possibly enrollees from there went to the Paliza CCC camp north of present-day Ponderosa. The University of New Mexico then leased from the CCC the buildings it had built at La Junta. The NYA camp at Battleship developed as a collaboration between UNM and NYA. The camp was to have a quota of one hundred boys for seasonal programs. The exact arrangement is not clear, but it seems the UNM archaeology field schools used the camp during certain parts of the year, and perhaps NYA directly employed some of the UNM students during the summer.

In 1938 the NYA/UNM collaboration began construction of "an administration building, seven barracks buildings, a mess hall and kitchen. Insofar as possible, the buildings will be constructed of native rock obtained near the site."[9] The seven old barracks buildings are still there, and at least two more cabins in the same style were added at some point. They are open-air cabins built on stone pillars and footings, made of rough-sawn timbers with rounded, bark-edged boards on the outside walls. The large mess hall and kitchen building is also still there today.

The barracks formed the primary bunk housing for the NYA/UNM camp and later for Camp Shaver. The experience of sheltering and sleeping in these open-air cabins under clear skies or during summer monsoons with lightning and rolling thunder up and down the canyons is an indelible memory for the thousands of kids and camp counselors who have spent summers here.

In addition to the historic open-air cabins and the mess hall, there are two large old buildings still being used on the far east end of Camp Shaver. One is a lavatory and shower building, called the Kybos during my summers at Camp Shaver in the late 1960s and early 1970s. (I'll explain the name later.) The other building, adjacent to the Kybos, is an impressive timber and stone lodge. The lodge has the look of a CCC-built structure, with a massive stone foundation and walls, and beautiful large vigas and latillas in the ceiling. I suspect the lodge was built during the CCC era (1933–1937), but it might have been part of the NYA/UNM construction period (1938–1944).

The CCC program was ended in 1942, and the NYA program in 1944. In 1945 the Albuquerque YMCA obtained permission from the Forest Service to use the camp and its buildings. The camp was named after William Shaver, who was a board member of the Albuquerque YMCA and was responsible for much of the organizing of the original Camp Shaver. He died in 1949 at age eighty-two. Today the YMCA owns about twenty-six acres of camp property on the west end. Some of the old buildings and open-air cabins on the east end are still on National Forest land (presumably under a special-use permit).

My personal memories of Camp Shaver are priceless. I recall first going there in 1965 or so, when we would visit my eldest brother, Jim, who worked as a dishwasher and cook's helper. During the summers he worked through the week at camp and came home for one night on the weekends. I recall that Jim and a couple of buddies bunked in an old building called the Crow's Nest (now gone). It was right behind the big mess hall and had a tall, two-story peaked roof, with their room on the top floor.

In 1969 I took over the job in the kitchen and mess hall, and Jim, who was a college student by that time, worked as a counselor that summer. I continued working at Shaver for three more summers as a combination

Figure 36.4. Camp Shaver director, counselors, nurse, dishwashers, and camp cooks on the front porch of the old mess hall, summer 1965. My brother Jim Swetnam is standing next to Rosa Rogers with arms crossed at lower right. Rosa was the camp cook and a long-time resident of the Jemez Valley. (Courtesy of Swetnam family)

dishwasher and cook's helper in the first couple of years and then later as a counselor.

It is fair to say that I literally grew up at Camp Shaver. Those were magical days of my fourteenth to seventeenth years. My family lived in Jemez Springs, and I attended Jemez Valley High School, but it was at Camp Shaver that I felt like I really met the world. The camp counselors were mostly college students from Albuquerque, while I was a younger and impressionable rural kid. The older campers tended to be my age or a little younger. These were years of the hippies in the Jemez (see chapter 39), the Vietnam War, the Apollo moon landings, marijuana, late-era Beatles and Led Zeppelin—and so much more.

One of the benefits of being a dishwasher/cook's helper at Camp Shaver was that I got to work during "girls camp." In those years, most of the one- and

Figure 36.5A. (*top left*) Cabin 7, where I was a camp counselor during the summer of 1972. The green window and door covers were added recently; the cabins were all open air in the 1970s. Photo taken in 2012.

Figure 36.5B. (*top right*) Campers and counselors on the bridge getting ready to fish, during Camp ARC, 1971. (Courtesy Mark Mexal)

Figure 35.5C. (*below*) The old mess hall, probably built in 1938 or 1939. We called the rim of Virgin Mesa in this view Cruiser Rock. Counselors took campers on long day or overnight hikes there. Photo taken 2012.

Chapter 36

two-week sessions at Camp Shaver were boys only. But for one two-week session, it was all girls, with women counselors, of course. So there I was, the only male teenager at camp, surrounded by more than a hundred girl counselors and campers near my age. Well, naturally, I experienced young romance.

Perhaps the most profound experience I had at Camp Shaver was during the only coed session of the summer, called Camp ARC. This was a special two-week session in cooperation with what was then called the Association for Retarded Children.[10] The campers were all intellectually disabled young people, although some were in their twenties, their thirties, or even older. During this session at least two counselors were assigned to each of the cabin groups, with one or two counselors in training (CITs) at each cabin. So Camp ARC sessions were very lively, with dozens of young women and men counselors.

The ARC campers were awesome. It was a huge eye-opener for me to see and experience this face of humanity. Generally, they were extremely loving and happy people. And there were some characters I will never forget—funny, sometimes difficult, sometimes easygoing, and often teaching you when you thought you were teaching them. Over the course of the two weeks, you could see the growth of friendships and loving care among the young counselors and ARC campers. I think everyone grew up a lot during those two-week ARC sessions.

And last, I mentioned that I would explain the name of the toilet/shower building, Kybo, more often referred to as the Kybos, plural, for the male and female sections. KYBO is a common acronym for "keep your bowels open." This slang is right in line with the fun and joking that happened day and night at this special place in the Jemez.

Place Names and Chronology of the Battleship Rock Campsite

Hemish ancestral farming and field houses (ca. 1300–1650)
Smith homestead and sawmill (ca. 1882–1912)
Camp Porter (Boy Scouts; ca. 1926–1933)
Battleship Rock Camp (UNM archaeology program; ca. 1930–1934)
La Junta Fly Camp and Camp Kearny (CCC; ca. 1935–1937)
Battleship Rock Camp (NYA/UNM; 1938–1944)
Camp Shaver (YMCA; 1945–present)

Figure 37.1. Turkey feathers applied to a length of yucca cordage in the Jemez style. (Courtesy Matthew Barbour)

CHAPTER 37

Turkeys in the Jemez

When I drive by Hummingbird Music Camp, I always think of the Higgins family, who started this summer camp in 1959 and still operate it, to the delight of countless kids and their parents. I am also reminded of Hollywood and turkeys. The many people, trailers, stages, and lights set up there in recent years for filming the *Big Sky* TV series are the latest Hollywood connection. An older connection goes back to the 1930s, when a Mr. Rowland owned this property and hosted a famous actor and turkeys.

Howard Leslie Rowland was born in 1895 in Philadelphia, Pennsylvania. His family were wealthy merchants and lived in a large mansion with servants. He was in the navy during World War I on a ship protecting convoys crossing the Atlantic. After the war he moved out west and attended the University of New Mexico. Throughout the 1920s he was a socialite in Albuquerque, with numerous mentions of him in newspaper social pages.[1]

By 1925 the newspaper articles refer to his Jemez Springs "ranch" and "hacienda," where he regularly held dances for local people and visitors to the three hotels operating here at that time. One of Rowland's guests in 1938 was Bruce Cabot, a Hollywood actor. Rowland had known Cabot from their days as fellow undergraduates at UNM. You might remember Cabot as a character actor in more than ten John Wayne films.[2]

The amusing story of Rowland's turkeys is told by Joseph Routledge, who grew up in the Jemez Mountains in the 1920s and 1930s. Here is a portion of the turkey tale:[3]

> There was a man who had a nice little place along with an orchard, on the Jemez River just north of what is now Hummingbird Music Camp. He was called a Gentleman Farmer, as he hired all his work to be done and never followed the walking plow. He was an expert at grafting fruit

trees of all kinds and had the best orchard in the canyon. No fruit of any kind grew further north than here because of the altitude and the cold. This Gentleman Farmer wore tailored clothing and was friendly but aloof. It was obvious he had money and the best of education. My father seemed to know him well. His name was Howard Roland [Rowland].

One summer he acquired a flock of white turkeys. We didn't know that turkeys came in white, and they were a great curiosity to us all. They were a little bigger than our wild turkeys and seemed to do well at Roland's place. That fall we had an early and wet snow in the high country and a lot of the wild birds migrated down out of the mountains to escape the snow and to harvest a bumper pinon crop. Now there is something good to eat: A roasted turkey that was fattened on pinons. I think you have already guessed that a wild gobbler and his flock descended on Mr. Roland and scooped up his white turkeys and headed for the rim. Thus began sightings of white turkeys that startled the locals to no end.

For the next 20 years, we saw white turkeys. The Fentons, Hofheins and Fettersons ate a white turkey once in a while, but we couldn't tell any difference in the taste. The white ones were the best targets. Even after the white ones were gone, there were mixed feather patterns everywhere. Mack Fenton killed a gobbler that weighed 24 pounds and was so big and fat he couldn't fly. The addition of the white bunch did not seem to add or subtract to the number of wild turkeys. We called them the Howard Rolands, or Jemez Springs turkeys.

Rowland died in 1958. He apparently never married or had children, and his estate went to probate court. The total value was a tidy sum, worth about $1.87 million in 2021 dollars. It seems the proceeds went to his niece Anne Rowland and various others.

In 1959 Lloyd and Wanda Higgins and a couple of partners bought the old Rowland ranch of twenty-five acres for $30,000. They established the Hummingbird Music Camp, which has been running for more than sixty years and now hosts Hollywood actors again, but not feathered turkeys to my knowledge.[4]

Chapter 37

As you feast on turkey during Thanksgiving, remember that turkeys and people have a long history in the Jemez. You may have seen wild turkeys here in the mountains. There is a fine Hemish petroglyph of a turkey on Virgin Mesa directly above Jemez Springs.

In 1934 a mummified child burial was found in Jemez Cave above Soda Dam. Carbon-14 dating revealed it was about 470 years old. This was a very special child. The child's body was wrapped in two deerskins and two turkey feather blankets. The blankets were made of thousands of downy body feathers of turkeys, all laboriously tied together with long, continuous strings made of yucca fibers.[5]

A recent research article on Puebloan turkey feather blankets noted that these were relatively common in Pueblo cultures, dating back to the Basket Maker periods in the early Common Era. These blankets were commonly used to keep children warm and were also used as burial objects. The authors estimate that a single blanket about the size of one found in Jemez Cave (about forty-eight by forty-eight inches) would require around 11,500 feathers from ten adult turkeys. One interpretation is that these blankets were made from feathers of a "Southwest domesticated turkey," not the same species as the wild Merriam's turkey present today in our landscape.[6]

The child burial bundle from Jemez Cave was finally returned to the Jemez Tribe about a decade ago, after residing—inappropriately—for many years in public exhibits at the Palace of the Governors and in archives of the Museum of New Mexico in Santa Fe.[7]

Figure 38.1. The church and steeple at Via Coeli, Jemez Springs.

CHAPTER 38

Servants of the Paraclete

Any landscape that is as culturally rich as the Jemez Mountains has many told and untold stories of human good and evil. This book has only scratched the surface of human history in the Jemez. One dark story that is only partly known by some, and unknown by others, is the role of the Servants of the Paraclete (SOP) in the sex abuse scandals of the Catholic Church here in the valley and around the world. Just ignoring this sad chapter might be the easiest thing to do. However, I believe the wisest path to reconciliation and avoiding the repetition of past injustice is in first telling and facing the truth—not burying it.

Father Gerald Michael Cushing Fitzgerald [1] was a cofounder of the SOP, whose motto was "Priests Help Priests." In 1947 he established a retreat here in Jemez Springs, a center where priests who were "spiritual casualties" could come for rest, prayer, and rehabilitation. Fitzgerald emerges from this history as a sympathetic character, but not fully blameless. He sincerely wanted to help troubled priests, and he devoted his life's work to that end. Initially, his plans came to fruition with the building of living facilities for wayward priests and a new church and rectory, named Via Coeli, directly across NM 4 from the ruins of the seventeenth-century San José de los Jémez Mission at Gíusewa. The twentieth-century church still stands today as the tallest human-built structure in the valley, with its steeple, minus the topping, a modernistic sculpture moved down to the adjacent ground.[2]

Since 1947 hundreds of priests with "problems" came to stay for months to years at Via Coeli and at the dispersed facility just north of Soda Dam known as Lourdes. As early as 1952 Fitzgerald realized that the church was sending more and more priests whose behavioral-psychological "problem" was pedophilia. They had committed heinous crimes of serial molestation of children from the parishes they had served in. Fitzgerald's original intention was to reform and rehabilitate alcoholic and depressive priests, and to return those who were healed back to parishes. But he believed that priests who had committed sexual crimes against children were not reformable or trustworthy and they should not be returned to parishes.

In a remarkable set of letters written to Catholic leadership during the 1950s and early 1960s, Fitzgerald argued consistently and vehemently that these abusive priests could not be rehabilitated; they should be defrocked (laicized) and sent to live on an island in an ocean, he said. He went so far as to find an island in the Caribbean and buy it. Here are a few quotes from these now publicly accessible letters to church leaders, including a letter to the Vatican:[3]

September 18, 1957

> Most Reverend and dear Archbishop
> Most dear Cofounder [probably Archbishop Edwin V. Byrne of Santa Fe, who was considered a cofounder of SOP]

May I beg Your Excellency to concur and approve of what I consider a very vital decision on our part—that for the sake of preventing scandal that might endanger the good name of Via Coeli we will not offer hospitality to men who have seduced or attempted to seduce little boys or girls? These men Your Excellency are devils and the wrath of God is upon them and if I were a Bishop I would tremble when I failed to report them to Rome for involuntary laycization [sic]. It is blasphemous to let them offer the Holy Sacrifice. If individual Bishops pressuring Your Excellency can say—Experience has taught us these men are too dangerous to the children of the Parish and neighborhood for us to be Justified in receiving them here. Your Excellency can if you wish say—you do not wish to interfere with the Rule-experience has dictated.

It is for this class of rattlesnake I have always wished the island retreat—but even an island is too good for these vipers of whom the Gentle Master said—it were better they had not been born—this is an indirect way of saying damned is it not?

> When I see the Holy Father I am going to speak of this class to His Holiness—they should be ipso facto reduced to lay Men when they act thus.
>
> In Spiritu,
> Fr. Gerald of the Holy Spirit, sP"

Repeatedly, in letter after letter to bishops and archbishops, Fitzgerald stated his strong opinions. Priests who were abusers of children were "vipers," he said; they should be defrocked and permanently removed from society. He used a particular biblical quote in several letters:[4]

> I am very much of the opinion that when a padre has fallen into the classification of this young man, he needs a very solid jolt to attempt (if this be possible) to achieve the realization of the gravity of his offence. Personally, I would want to spend the rest of any life on my knees asking God's Mercy, for I know, no more terrible threat than the words of Our Lord: "those who tamper with the innocence of the innocents—it were better if they had never been born."
>
> Men who sin repeatedly with little children certainly fall under the classification of those who "it were better had they not been born." He will hurt the Church; and he will hurt your Community. Moreover, as a layman, the civil authorities will make short work of his activity and place him in the protective custody that his type merit. Otherwise, sooner or later he will kill or be killed. There comes a time when an individual who has repeatedly abused his priesthood should get retributive action.
> . . . As there are many little children in this Canyon, where I am the shepherd of souls, I could not in conscience consider receiving him here. Very cordially in Christ, Father Gerald Fitzgerald, Servant General
>
> I trust that this does not seem too severe. But I have my own soul to save, and I do not dare recommend such men for the cure animarum ["care of souls"].

In 1963 Fitzgerald had an audience with Pope Paul VI at the Vatican, and he laid out his concerns about the pedophilia crisis in the church. The pope requested a follow-up report, which Fitzgerald provided. Unfortunately, his warnings were ignored. Church leaders dismissed his pleadings as an "emotional" overreaction and his island isolation idea as bizarre.

Fitzgerald distrusted psychiatric treatment for pedophilia, but others in the Church disagreed and argued that some pedophilic priests treated with modern methods could be reintegrated into parishes. By 1965 he had lost this argument, and he was elbowed out of leadership of the SOP. He no longer lived in Jemez Springs after that. He died in 1969 and is buried here in the Paraclete cemetery.[5]

After the Fitzgerald era at Via Coeli ended in late 1965, the SOP embraced psychology, psychiatry, and medicinal treatment for priests accused of serial sex abuse of children. They built a psychiatric hospital on the SOP property in Jemez Springs. For more than twenty-five years after that, dozens of abusive priests were treated at the SOP, and subsequently many of them were sent back to parishes.

Tragically, the treatments were ineffective in many of the SOP pedophilia cases, despite unsubstantiated claims at the time and later by mental health practitioners who had worked there. No peer-reviewed studies on the SOP treatment effectiveness have been published, so we don't know the actual rate of recidivism by priests treated at the hospital. But we do know, from many investigative news articles, court cases, and settled lawsuits, that many priests who passed through the SOP rehabilitation programs went on to continue their sexual abuse of hundreds of children.[6]

Here are just two examples among the many priests treated at the SOP and then recycled back to parishes. James Porter was one of the most prolific pedophile priests treated at the SOP. He was sent to Via Coeli multiple times and repeatedly returned to parishes, including parishes in New Mexico. Porter admitted to sexually abusing more than one hundred children during his career. He was convicted of child molestation in 1993 and sentenced to eighteen years in prison.[7] Michael Stephen Baker has been called by authorities one of the Los Angeles archdiocese's most prolific child molesters, with estimates of between twenty-three and twenty-eight victims, one of whom was as young as age five. Baker spent time in treatment at the SOP in Jemez Springs in 1987, and his convictions for sexual abuse of multiple children occurred after this visit. In 2007 Baker pleaded guilty to the sexual abuse of two young brothers and was sentenced to ten years and four months in prison.[8]

The treatment at SOP and the recycling of pedophile priests back to parishes continued into the 1990s. But by 1995 the scandals and lawsuits were at a landslide scale, so around that time the church said the SOP facilities in Jemez Springs were no longer to be used to treat priests accused of child sexual abuse.[9] The SOP still has other facilities housing sex abuser priests, including perhaps their last major US stronghold near St. Louis, Missouri. An investigative documentary film about the SOP in both Jemez Springs and St. Louis was recently aired on cable TV and can be viewed online.[10]

A particular case illustrates how far off the rails the leadership at the SOP in Jemez Springs became both before and after Fitzgerald was ousted. In 1960 Father John Bernard Feit was a parish priest in McAllen, Texas. A twenty-six-year-old teacher and former beauty queen from town went missing, and later her body was found. She had been raped and murdered. Feit was a prime suspect based on limited evidence linking him to the crime, but no charges were filed. Two months later Feit was charged with sexually assaulting another woman, but that charge was reduced in a plea deal; Feit pled guilty and paid a five hundred dollar fine. Feit was then sent away by the church, and he eventually ended up at the SOP in Jemez Springs.[11]

Apparently, Feit was persuasive and competent in some ways, because over time he became a staff member at the SOP while Fitzgerald was still in charge and later a top administrator there. He was one of the key leaders at Via Coeli during the period after 1965 when multiple child abuser priests were sent back into parishes. The parishes in Jemez Springs and the broader Jemez Valley were not exempt from the recycling of abusive priests—and their abuses—during this era of the 1960s to 1980s. And Feit wasn't the only sex abuser channeled into a staff position at an SOP facility.[12]

Feit left the priesthood in 1971 and later married and had three children. Finally, in the early 2000s, two priests came forward and testified that Feit had confessed to them very soon after the murder that he had killed the young woman. The case lingered in Texas for nearly two decades, but eventually Feit was charged. He was tried and found guilty of murder in 2017. By that time he was eighty-four years old. He died of natural causes in a Texas prison in 2020.

The SOP no longer brings priests to the valley, and its properties have mostly been sold. The old psychiatric hospital on the west side of the river is now owned by the National Park Service, and it is used as the headquarters of the Valles Caldera National Preserve. The largest SOP landholding, about two thousand acres surrounding the village of Jemez Springs, was recently purchased by the Forest Service. The church was forced into selling these properties to help meet debts incurred from dozens of lawsuits relating to sexual abuse by priests. US taxpayers are now helping pay this debt, in return for lands the SOP bought in 1947. Those lands were formerly part of the San Diego Land Grant, swindled away from the valley's Hispano heirs by unscrupulous men in the early 1900s (see chapter 11).

I have no specific solutions or recommendations for going forward. However, I wonder if the people of Jemez Springs should reconsider the name of the central village park (Fitzgerald), and I wonder if the only named boulevard in the village (Mooney) should commemorate administrators of the Servants of the Paraclete.[13]

CHAPTER 39

Hippies and Hot Springs in the Jemez

The Jemez Mountains have always attracted interesting characters. The springs are a powerful lure for visitors to come and soak in hot mineral waters while enjoying the beauty of our canyons and forests. Growing up in Jemez Springs in the late 1960s and early 1970s, I witnessed a remarkable subculture of hot spring visitors—the hippies! As everyone knows, the hippie movement exploded in the 1960s, along with protests against the Vietnam War; the proliferation of various illicit drugs, especially marijuana; and sexual liberation. Many young people across the United States and the world embraced some form of the philosophy of former Harvard lecturer and LSD guru Timothy Leary, reflected in his slogan "Turn on, tune in, and drop out."

The most memorable summers of the hippies in the Jemez were from 1967 to 1969, when hundreds of members of the Hog Farm and other communes arrived in psychedelic-painted school buses. What a sight that was! One of the leaders of the Hog Farm was Wavy Gravy (Hugh Romney). He became famous after his colorful announcing and antics at the Woodstock Music Festival in New York in August 1969. The Hog Farm notoriety brought so many hippies to its Taos commune that by the winter of 1969 it had a crisis in feeding and housing all the young longhairs.

I was told that members of the Hog Farm camped in the Jemez along San Antonio Creek, near the hot spring up there. Some hippies stayed the whole summer, including at San Antonio and Spence Hot Springs, and McCauley (Abousleman) Warm Springs. This became an irritation for locals and other visitors wanting to use the springs without being exposed to nudity, overcrowding, and drugs. The hippies also tended to trash the areas around the springs with overuse.

Figure 39.1. Members of the Hog Farm in Spence Springs, Jemez Mountains, 1967. This photo is by Lisa Law, one of the founders of the Hog Farm commune in Taos. She was a prolific photographer documenting the days and people of the counterculture. (Courtesy Lisa Law and National Museum of American History, Smithsonian Institution, Catalog No. 1998.0139.105)

In addition to the fabulous painted buses, I recall seeing hippies all around the village, including in the mercantile, in Los Ojos Bar, and walking up and down NM 4. One character stands out in my memory. This guy called himself Ulysses S. Grant, claiming he was reincarnated from the Civil War general and president. His real name was Donald Waskey. Grant/Waskey often traveled through Jemez Springs riding bareback on a large white horse he called Charger. Usually, one to a few women and children followed behind on foot.

In 1970 Waskey led a commune in Placitas known as the Lower Farm. He ran for governor of New Mexico as a write-in candidate in November 1970 and received six votes. In December of that year, he got into a dispute with other commune members, and he fatally shot two men and wounded

Figure 39.2. Hog Farmers and friends on a psychedelic-painted school bus, 1968. "The Road Hog, El Rito 4th of July parade, New Mexico, Wavy's Bus." (Courtesy Lisa Law and National Museum of American History, Smithsonian Institution, Catalog No. 1998.0139.127)

a third. Waskey, his wife, and their young child then disappeared. Despite a nationwide manhunt, he was never found alive.[1]

Eighteen years later, in 1988, the bodies of Waskey and his wife were finally found on a remote Idaho farm. They had been killed execution-style. At the farm authorities found hundreds of marijuana plants in what was described as the largest pot-growing operation in the state at that time. Nobody was ever arrested for the murders.[2]

Fortunately, there was no serious violence associated with the hippies in the Jemez during those turbulent years of the 1960s and 1970s, to my knowledge. However, other conflicts posed a problem for forest rangers. Complaints about dozens of nude hippies visible from the highway, and overcrowding and overstaying at the hot springs, eventually led the Forest Service to issue an official proclamation and rules in 1975. Some days of the week at Spence Springs were designated "bathing suits not required" (Sunday to Wednesday), and other days they were required (Thursday to Saturday).

My friend and longtime Jemez Valley resident Cathy Stephenson was a law enforcement officer for the Forest Service at the time. She was ordered to patrol the hot springs and issue tickets to people bathing naked on the wrong days. She told me,

> Every one of the bathers denied having ID, and it wasn't possible to tie them to a vehicle—since the vehicles were parked across the canyon and many had hitchhiked anyway. So I would say, "Everyone out of the pool and give me your ID," and they would all get out of the pool and deny having ID. And I would write up the tickets with whatever fake names and addresses they gave me. I recall some of the guys would wag their equipment at me and quip—"search me lady." They would all be back in the water by the time I got back to my truck. Not a single ticket was ever paid.... Before, during, and after the regulations, our ranger district's policy was to tell everyone who inquired about Spence that there would be nude people there, despite any ruling to the contrary. Short of twenty-four-hour guards making them keep their clothes on, there was simply no way to ensure a person wouldn't encounter nude bathers.[3]

My father, Fred Swetnam, was the first Jemez District ranger to try to manage the onslaught of hundreds of hippies at the hot springs in the late 1960s. He was a WWII veteran and an old-school ranger in some ways and a modern one in others. He considered the hippies a "subculture." He also tried to be fair and even-handed with them. The following newspaper article from March 1970 quotes him. The article captures the conflicts and the Forest Service's dilemma at that time:[4]

> Jemez Springs—US Forest Service officials here are working to avoid an almost inevitable clash between hippies and summer visitors to the colorful Santa Fe National Forest Canyon, long a favorite escape from Albuquerque's hot summers. If such a confrontation is avoided it will largely be through the efforts of Fred R. Swetnam, Jemez District Ranger and his staff. Swetnam is trying to evolve a policy which will permit both groups to use the recreation area without conflict.

Chapter 39

Source of the trouble is mixed nude bathing by transient hippies within sight of traffic along NM4, north of Jemez Springs. On some occasions from 200 to 300 hippies have been observed sunbathing in the nude near the spring which could create a difficult law enforcement problem.

Several courses of action are being considered. Use of state indecent exposure laws has not been overly successful. Swetnam believes stringent enforcement of Forest Service sanitation laws would be more helpful in control of the problem. As a final resort the Jemez Forest Service might dynamite a hot spring pool which has been a favorite hippie gathering place. "Once the hippie problem has abated, we could restore the pool," he said. The suggestion was made more than a year ago but was strongly opposed by Jemez Springs residents. "Now they are beginning to think it might not be such a bad idea after all," the Ranger said. But he admits that the last resort action would amount to surrender to the hippies and would not successfully resolve the problem.

"Mainly we're trying to live and let live and preclude clashes between the two cultures," Swetnam said. Residents of the community itself have been kindly toward hippies on their periodic visits. Townspeople are not particularly happy about what they hear is occurring at hot springs in the canyon, but they have tended to assist hippies in trouble rather than trying to run them off.

Swetnam who wryly observed that, "nothing in my training, experience, or regulations quite prepared me to cope with the hippie situation," has negotiated agreements with three identifiable groups. The agreement reads in part, "The Forest Service recognizes your people as ordinary citizens. You are entitled to equal privileges and benefits under the law. You have no additional privileges over other citizens, nor do other citizens have privileges that you do not have. All are equal in the eyes of the law and with the use of the hot springs. . . .

"We recognize your ability to communicate with your people We have great faith in this ability. It is our desire that you communicate with your segment of society and help us govern their actions in that clashes with other segments may be avoided."

> **SPENCE HOT SPRINGS**
> **BATHING SCHEDULE**
> ▪ BATHING SUITS NOT REQUIRED ▪
> SUNDAY MONDAY TUESDAY
> &
> WEDNESDAY
> ▪ BATHING SUITS REQUIRED ▪
> THURSDAY FRIDAY
> &
> SATURDAY

Figure 39.3. Forest Service sign posted at Spence Springs in the 1970s. (Courtesy Cathy Stephenson)

Swetnam said that the Forest Service does not challenge anyone on "your thoughts, your existence, your religious beliefs, your rights as Americans, but we believe that actions of your group infringe upon the rights and privileges of the other segments of society. We are not saying which segment is bigger or which is the right segment. To avoid serious confrontation, we are trying to keep the two apart. Ours is a system of majority rule, coupled with protection of the minority—this is the American way."

Swetnam asked for "some demonstrated acts of good faith on your part. We seek only to lessen the tension between the groups." He said points of sensitivity and tension involve camping in no-camping areas, disregard of rules governing use of hot spring pools, and leaving the pool area in the nude and standing or sitting on rocks in view of the road while naked.

He told the hippie's representatives, including Ulysses S. Grant and a leader of the "Hog Farm Lodge," that "we are making a sincere attempt to communicate with your segment of society. If this cannot be successfully done it is only fact that more law enforcement practices must follow."

Law enforcement practices used thus far include charges of indecent exposure brought under state statute with cooperation of the New Mexico State police. There has been one conviction on that charge, and conviction on a charge of starting a forest fire with burning underwear. In all, charges have been brought against 12 persons since last summer. One of the difficulties in prosecuting under the state indecent exposure statute is the law's ambiguity and reluctance on the part of complaining witnesses to testify.

Chance encounters bring persons to the Ranger station "boiling mad, but by the time they get to court the accused hippie has apologized for offending anyone and the complaining witness has cooled off. That often is the end of that," Swetnam said. A "hippie patrol" which includes Ranger station personnel visit hippie camps and "encourage them to keep their clothes on, keep their camps clean, and observe Forest Service regulations." Members of the patrol point out that most hippies "keep their camps cleaner" than many "average" campers. "We don't have much trouble with them." Swetnam describes himself as "a man walking a tightrope, between the middle-class church-going Americans whose sensibilities are offended, and the longhairs lacking in the modesty which sets us apart from the savages and animals."

Presence of hippies in the forest which has a "two week stay limit" is driving away some of the other outside visitors, Swetnam said. "The minority has won up here. Now we have to decide if we can force the majority back on them. Forest lands are designated for many uses, and use by the hippies is just another use. The Forest Service is caught in the middle of the problem in which we are trying to avoid serious confrontations between the dominant culture and the subculture. We have to resolve the conflict with the Forest Service's philosophy of the greatest good for the greatest number involved in mind."

When I posted this fifty-year-old newspaper article on the "Jemez Chat" Facebook page a couple of years ago, there were several interesting replies. One resident recalled that the San Antonio Hot Springs were blasted by dynamite and people were upset at the time. However, Cathy Stephenson is confident that the San Antonio Springs were never blasted. I first visited that spring in 1976 and it was about the same as it is today. Another resident posted a photo of an old sign from a hot spring trail that someone had given to her. It said, "Spence Spring. Clothing Optional. Act Normal. Nudists are People Too."

A little confession: When I was a teenager working at Camp Shaver during the summers of 1969 to 1972, some of us camp counselors went to Spence Springs, soaked with the hippies, and indulged in smoking certain illicit substances. Fortunately, the hippies "negotiating" with my dad at the time didn't know who I was, and I didn't tell my dad about my hippie hot spring adventures either—until decades later.

When I did tell my dad about dipping with the hippies, he found it amusing, as he often told the Jemez hippie stories with humor when we visited him in his later years. (He died in 2006.) He would pull out of his files hilarious editorial cartoons that had appeared in New Mexico newspapers. One of the cartoons showed a forest ranger staring at naked figures on a distant mountainside using binoculars. The ranger's eyes were bulging out the ends of the binoculars!

CHAPTER 40

Sierra de los Valles

A paradoxical but lyrical place name for the Jemez Mountains is Sierra de los Valles ("mountains of the valleys").[1] On old maps this name is applied specifically to the eastern rim of the Valles Caldera, but its general meaning is descriptive of the whole mountain range. Satellite views from orbit show the Jemez as a distinct, circular ring of mountains, dotted with peaks, parks, and valleys within and around the circle. The Nacimiento ("birth" or "origin") Mountains are a straight-edge line of peaks just outside the ring of mountains on the west side. A high plateau forming the San Pedro Parks anchors the northwest corner of the mountain mass.

The Jemez landscape encompasses many places and many stories. This book has described only a small fraction of those places and has told only a few of its stories. Some of the stories are from the memories of people told by oral tradition or written in books and articles. Other stories are told by trees and rocks in languages that we can only partly comprehend. Still, if we listen closely, there is much to be learned. As Keith Basso recounts in his fabulous book *Wisdom Sits in Places*, localities can represent memories, and they can also teach lessons.[2]

Basso describes knowledge and wisdom in stories connected to Western Apache landscapes in Arizona. Many places there have old Apache names and stories associated with them. Many of the stories contain an object lesson, a nugget of wisdom to be taught. The landscape is then a mnemonic for recalling good or bad, right or wrong, or what is simply a sense of place. I don't know if all Hemish place names in these mountains have similar origins or functions, but I suspect many of them do.[3]

If you live within a landscape long enough, travel around it, have good and bad experiences in many different places, and listen to the stories told about each of the nooks and crannies of that landscape, then you can build within your mind a "memory palace." This is an ancient mnemonic

Figure 40.1. An oblique view of the Jemez Mountains, with Rio Jemez joining the Rio Grande at the bottom center and the Rio Chama joining the Rio Grande at the upper right.

device going back to the Greeks. This memory tool is used for recalling many things in a sequence, and it depends upon the connections established between memories of places.[4]

Imagine a grand building, a palace, where you intimately know the spatial arrangement of the floors, rooms, and hallways. Then you can begin to put objects, stories, or people within each of those places—in your mind. The spatial memory of the palace provides a mnemonic, a framework for recall of those things. Memory champions use the memory palace technique to memorize decks of shuffled cards in order, placing each card, in their mind, in a set sequence of locations. Then they walk through the rooms and hallways of their palace in a set sequence, recalling each card in the correct order.[5]

I think of the Jemez Mountains as a grand memory palace. It is an intricate spatial landscape with high mesas; deep canyons; valley bottoms;

Chapter 40

wide grassy valles and parks surrounded by mountains; massive cliff faces; hot, warm, and cold springs bubbling out of hillsides; travertine and sulphur deposits; ancient trees; village, house, and church ruins; and much, much more. But this memory palace is not simply a device for memorizing facts or history.

Seeing and visiting these places also can lead to the recall of people and events related to those locations and the making of connections and new discoveries. Sometimes those stories are surprising, strange, or awesome; funny, tragic, or maddening. Sometimes they impart knowledge—and sometimes even wisdom. Places on the landscape remind us of deep time stories told of and by rocks and trees and near time stories of people. The stories overlap and stitch together, forming a diverse quilt of memories.

There are memories of volcanic eruptions, floods of lava and water, lakes, and pyroclastic flows, travertine mounding, wildfires, rivers flowing, and eons of erosion. There are also memories of people in their villages with their feet moving in time to pounding drums, of cornfields, raiders, horses, priests and soldiers, cannons and muskets booming. And there are memories of people gathered in a church, singing hymns accompanied by a "wheezy harmonium," making a "joyful noise."

When I was a boy, one of my favorite adventures was a day hike on the Padre Alonzo Trail from the old mission church up to the top of Cat Mesa or up the west-side trail to Virgin Mesa. I'll never forget the ecstatic feeling of topping out on the rim where the trails entrench in the bedrock. A few steps over to the cliff edge, and there are the grandest views of the Jemez Valley. The canyon bottom is more than a thousand feet below. The view to the north is of Redondo Peak, Wâavêmâ, rising above Banco Bonito and Battleship Rock. To the south is the long valley and winding cottonwood bosque, all the way down to the red rocks at the junction of the rios Jemez and Guadalupe.

Standing on the cliff edge of Virgin or Cat Mesa is thrilling, partly because of the vast open spaces and the slight danger. But it is also wonderous because the whole landscape is laid out within view, and I am reminded of so many places and people, memories layered and superimposed in time. Hopefully, the stories in this book will now find places within your Jemez memory palace, increasing your wonder and enjoyment of this special place.

The Jemez Mountains

NOTES

CHAPTER 1

1. Elliott, *Large Pueblo Sites*; Kulisheck, "The Archaeology of Pueblo Population Change."
2. Liebmann et al., "Native American Depopulation."
3. Kulisheck et al., "Agricultural Intensification."
4. Bloom, "West Jemez Culture Area."
5. In chapter 7 J. K. Livingston describes his family pack trip into the Jemez in 1886, including a ride up the Padre Alzono Trail to visit Monument Canyon.
6. Vallecitos Viejo is now known as Ponderosa, and the mesa to the east is Borrego Mesa. The big village ruin up on that mesa is called Pejunkwa and was possibly thirteen hundred rooms in size, according to the archaeologist Mike Elliott. It is unusual for a Jemez ancestral village in that it seems it contained both stone and adobe buildings rather than mainly stone, as most of the other big ruins on the mesas to the west. Bloom was probably incorrect in stating that these sites "were occupied by the ancient Cochiteño." Hemish people and archaeologists (including Elliott) have identified the large mapped ruin sites on this part of Borrego as ancestral Jemez villages.
7. Julyan, *Place Names of New Mexico*. Julyan writes, "Canova (Rio Arriba, NE of Espanola, on the west side of the Rio Grande, opposite Velarde). Tiny, inhabited community whose Spanish name means 'water trough, sluice,' doubtless a reference to the extensive irrigation in the area."

CHAPTER 2

1. Elliott, National Historic Landmark nomination form, 23. Elliott describes impressions of the church:

    In 1634 Benavides described the church and convento as "very sumptuous." ... In 1915 L. Bradford Prince, former New Mexico governor, declared "the most beautiful ruin in New Mexico, beyond all compare, is that of the old

Notes

Mission Church at Jemez.... In 1923 historian Lansing Bloom included Jemez in his list of the "three finest examples of the early missions of New Mexico."... In the late 1920s journalist and historian Earl R. Forrest called the church "among the most picturesque in all New Mexico. This was a massive building; with walls eight feet thick it has stood throughout the ages, and in its time must have been a beautiful edifice."

2. Sando, *Nee Hemish*.
3. Sando, 107.
4. Sando is referring here to the *Relaciones* document, which quotes Salmerón from his report written in 1626.
5. Matthew Barbour, manager of the Jemez Historic Site, wrote about a 1623 uprising by the Hemish that was apparently the cause of the burning of the church only two years after it was first completed. This uprising and subsequent military response by the Spanish led to much upheaval in the Jemez Mountains. It is unclear (to me) how Zárate Salmerón escaped this uprising, but according to Barbour and Sando, the next priest, Arvide, and the Spanish forced the Hemish people to resettle at Gíusewa and reconstruct the church after 1626 or 1628. Matthew Barbour, "The Pueblo Revolt of 1623," *Jemez Thunder*, October 1, 2015, 29.
6. Elliott, National Historic Landmark nomination form, 5, quoting a report by James Ivey.
7. Sando, *Nee Hemish*, 106.
8. Tree-ring dates of the house built within the convent in Elliott, National Historic Landmark nomination form, 34–35. Dates and usage of the church and kiva construction in the convent area in Robinson et al., *Tree-ring Dates from New Mexico*.
9. Mike Elliott kindly investigated a question about the staircase: Was it the original or a replacement built by the Civilian Conservation Corps (CCC)? He sent to me an excerpt of a 1910 report by Edgar Hewitt describing the interior of the tower, with empty slots in the walls where steps once existed, and also a photo from 1922 by John H. D. Blanke that is almost the same view shown here, but without the steps and showing the empty sockets in the walls. The loophole in the north wall is just visible in the photo in this chapter and is seen completely in the 1920s photo. It is remarkable, I think, that the CCC went to the trouble and effort of rebuilding that stairway in the 1930s. That required hoisting the

large, heavy central pole and wooden steps up and into the tower via the narrow slot on the south side and installing them in the tower pit and walls.
10. Photos of the frescos are in Elliott, National Historic Landmark nomination form. The captions are "The plaster walls of the church nave were colorfully decorated with elaborate designs" and "Art student Marguerite Tew of the 1921–22 expedition produced copies of the nave designs in color."
11. Bandelier visited the old mission twice, in 1888 and 1891. In the nomination form (36–37), Elliott quotes Bandelier: "There are quite a number of Spanish houses about the church, and the arroyo is lined with solid walls, to protect the houses." Elliott also writes, "His journal [Bandelier] recounts a visit with Dr. Shields, who showed him through the ruins. Bandelier listed artifacts present, including pieces of pottery (white with black lines), a stone axe of red porphyry or granite, an iron knife, and metates and manos of lava and sandstone. He reported the ceiling of the church featured 'plaited willow work.'"

CHAPTER 3

1. The name for Guadalupe Mesa in Towa, Munstiashinkiokwa, is given by Paul Tosa in a short article about the Battle of Astialakwa in the Jemez Pueblo newsletter the *Walatowan*, October–November 2021.
2. Sando, *Nee Hemish*; Liebmann, *Revolt*; Roberts, *Pueblo Revolt*.
3. Flint and Flint, "Bartolome Ojeda"; Kessell, *A Long Time Coming*.
4. Those of mixed heritage, with combinations of Pueblo, Spanish, and African descent, including second generations, were labeled various terms, including *ladino*, *mestizo*, and *coyote*. A number of current or former African slaves were among the New Mexico colonists in the late sixteenth and seventeenth centuries. The term *coyote* may have been used interchangeably with *mestizo* in New Mexico (see Weber, *What Caused the Pueblo Revolt?*), but Liebmann says it was a term used for "those born of Native American and African union" (54). Katzew (*Casta Paintings*) says that *coyote* (or *coyota*) refers to a person born of one mestizo (mixed Spanish and Indigenous) parent and one Indigenous parent. In addition to Ojeda, other mestizo or coyote leaders who were among the rebels during the 1680 revolt became warrior allies of the Spanish after 1692.
5. See Liebmann, *Revolt*, 167, indicating that others have questioned the accuracy of this version of the 1689 battle at Zia. Other stories of the battle say

Notes

it occurred while most Zia warriors were away from the pueblo and that the number of deaths was exaggerated.

6. Whether Ojeda's conversion to the Spanish cause derived from a revelation during his recovery from the Zia battle of 1689 or he was attempting to minimize impacts of the reconquest by joining the Spanish is a subject of historical speculation. Although he was a key war captain of Pueblo auxiliaries fighting for the Spanish during the reconquest, at times he served to spare Pueblo fighters from execution. Flint and Flint write, "In just a few months, Ojeda had transformed from a violent foe of the Spanish provincial government to an agent of its reestablishment. Indeed, from the time of his recuperation, he seems to have worked consistently on behalf of Spanish reoccupation of the Rio Grande region. Presumably, his unanticipated and remarkable cure under Franciscan care had created an epiphany in his mind. Or it may have been, as Kessell has written, that 'to save at least some of his people, he resolved to ease the Spaniards' inevitable reentry into the Pueblo world.'"

7. Flint and Flint 2009, "Bartolome Ojeda."

8. Joshua Madalena, a historian of Jemez Pueblo, tells this story of the infiltrator spies who turned on the defenders at Astialakwa; quoted in Liebmann, *Revolt*.

9. Liebmann describes the ambiguity of the saint name applied to the image on the cliff:

> Images of both San Diego (on the east side of the peñol) and the Virgin (on the west side) are said to be visible on the cliffs of the mesa today and are still venerated by the contemporary people of Jemez (as well as the non-Native residents of the Jemez Valley). Confusingly, the image purported to represent San Diego (St. Didacus, patron of Franciscan laity) actually seems to be that of Santiago (St. James, patron of the Spanish empire, soldiers and the reconquista). The likeness emblazoned on the side of the cliff (comprised of a series of cracks, variations in the color of the stone, and water stains) resembles a bearded man in profile standing above a distinctively equine shape. Colonial representations of Santiago commonly depicted him with flowing white beard astride a magnificent steed ... thus the apparition on the peñol is probably more appropriately termed that of Santiago, not San Diego. Significantly, the battle at Astialakwa occurred just one day prior to the feast of Santiago

(July 25) as well. "Doubtless it is his patronage and intercession," opined Vargas, "since it was the eve of his glorious day, that played a large part in the triumph."

10. Sando, *Nee Hemish*, 110.
11. A fanega is defined in different ways in old Spanish usage, depending on the region. The equivalent could be as high as 1.6 bushels, and the weight of a bushel of shelled or unshelled corn would depend on how dry it was. Liebmann (205) states that this quantity was equivalent to 58,800 pounds of corn, which would be more than twenty-nine tons. Perhaps this would be the volume and weight of unshelled corn on the cob? In any case, this was a huge haul of corn. Hundreds of sheep and other livestock of the Hemish were also seized, and most were given to the Pueblo auxiliaries of the Spanish.
12. Liebmann (220) recounts the final coalescing of Patokwa and Walatowa in the mid-1700s.
13. The name of this petroglyph—Dancing Coyotes—was told by Barnabe Romero of Jemez Pueblo to Cathy Stephenson in the 1970s. Both Romero and Stephenson were US Forest Service employees documenting archaeological sites on the Jemez Ranger District. Stephenson shared the name and location with me. The great antiquity of this petroglyph is evident in the extensive growth of several species of very slow-growing lichen on the carvings. A Hemish elder told me that this petroglyph is related to a special dance once performed at the village of Potokwa. He said it is not directly related to the battle of 1694 or 1696. Still, I wonder what the coyotes may signify in the ancient dance and as carved on the boulder. Interpretation of the meaning of ancient pictographs and petroglyphs is a fraught endeavor, prone to just-so story explanations. Nevertheless, I offer some interesting observations that may be coincidental or alternative meanings of the carved coyotes: (1) Mixed-blood people in New Mexico were sometimes called coyotes. Bartolomé de Ojeda was a coyote. Other warriors fighting for the Spanish during the reconquest were also coyotes. (2) Sando (110) writes, "Oral historians of the Jemez used to talk about this period [July 1694, when the Spanish and their allies arrived for battle] as the time when the pack-carrying burros could be heard braying like coyotes." (3) The coyote figure in North American tribal mythologies is sometimes known as the trickster, one who regularly fools and double crosses. In this context, note the Pueblo infiltrator spies who turned on the Astialakwa

defenders at a crucial moment. (4) In a 1942 *Desert* magazine article, a Navajo medicine man said that Ma'iidesgizh (Coyote) was the Navajo name for the saddle between Guadalupe Mesa and Virgin Mesa and also the name of a Navajo clan associated with people at the battle.

CHAPTER 4

1. Quoted in Benavides, *Fray Alonso de Benavides Revised Memorial.*
2. Quoted in Benavides.
3. Simpson, *Navaho Expedition,* 15–18.
4. Simpson, 18.
5. Elston, *Summary of the Mineral Resources.*
6. Romer, "Vertebrate Fauna of the New Mexico Permian."
7. Lucas et al., "Lithostratigraphy."

CHAPTER 5

1. Fowler, *Western Photographs of John K. Hillers.* The New Mexico History Museum labels this photo "circa 1880." I think it was from 1879. Fowler identifies the man in the arroyo as Stevenson. See the photo caption for figure 5.2.
2. "James Stevenson," *Science* 12, no. 288 (1888): 63–64; Miller, *Matilda Coxe Stevenson.*
3. Fowler, *Western Photographs.*
4. Miller, *Matilda Coxe Stevenson.*
5. Holmes, *Random Records.*
6. Miller, *Matilda Coxe Stevenson.*
7. See Nelson, "Favorite Sun."

CHAPTER 6

1. Bandelier, "The 'Montezuma' of the Pueblo Indians."
2. Anaya et al., *Aztlán.*
3. Bandelier, "The 'Montezuma' of the Pueblo Indians."

4. Read, "The Last Word on Montezuma."
5. Ritch. *Aztlán*. There are multiple retellings of various versions of the Montezuma and New Mexico myths. Some of the most lurid involve a "sacred fire" and a guardian serpent in a kiva at Pecos Pueblo. An example is a story told by Matthew C. Field in 1839 and recounted in Kessell, *Kiva, Cross & Crown*. Also, J. M. Scanland's "Theft of the Sacred Fire" was a tall tale published multiple times in 1902–1903 newspapers, along with pen-and-ink illustrations. This wild version, with a sacred fire element of the myth, ends with the eruption of a Jemez volcano.
6. Gutiérrez, "Aztlán, Montezuma, and New Mexico."
7. Read, "The Last Word on Montezuma."
8. Crown and Hurst, "Evidence of Cacao Use."
9. Fowler, "Some Lexical Clues."
10. White, "Pueblo of Sia," 20.

CHAPTER 7

1. J. K. Livingston, "Visit to the Valles, Jemez Springs, and Monument Cañon, a Veritable Land of Modern Wonders," *Santa Fe New Mexican*, July 18, 1886.
2. Wheeler, *Explorations and Surveys*.
3. Wheeler.

CHAPTER 8

1. deBuys, *Seeing Things Whole*.
2. *Stegner*, Beyond the Hundredth Meridian.
3. John Wesley Powell, "Irrigation of Arid Lands," statement and testimony to US Senate committee, February 15, 1889; John Wesley Powell, "Reclamation of Arid Lands by Irrigation," statement and testimony to Committee on Territories, US House of Representatives, February 16, 1889.
4. Holmes, *Random Records*, 259–75.
5. Holmes, "Notes on the Antiquities of Jemez Valley."
6. Holmes, *Random Records*, 94.
7. Holmes, *Random Records*.

CHAPTER 9

1. Ghattas, *Los Árabes of New Mexico*.
2. *Las Vegas Daily Optic*, May 29, 1897.
3. Abousleman Family Tree, Ancestry.com, accessed July 15, 2023.
4. *Albuquerque Journal*, December 31, 1904.
5. Swetnam, "Tree-Ring Dates from the Hot Sulphur Bath House."
6. "Judt Sells His Jemez Hot Springs," *Albuquerque Citizen*, July 29, 1905.
7. "Fire at Jemez Hot Springs," *Santa Fe New Mexican*, May 21, 1900. This report says the fire was in the home of the Abouslemans, which was attached to the store, and they stopped the fire before much damage was done. According to "Fire at Jemez Hot Springs," *Albuquerque Daily Citizen*, March 31, 1903, two stores were destroyed and insurance "almost" covered the losses. The May 8, 1921, *Albuquerque Morning Journal* reported, "The store of Moses Abousleman, at Jemez pueblo, was destroyed by fire Sunday night."
8. "Safe at Jemez Springs Robbed of a Thousand, Moses Abousleman Victim of Bold Thief, Suspect Arrested at Indian Village of Zia," *Albuquerque Morning Journal*, July 22, 1906; "Man Is Held on Attempt to Commit Robbery," *Albuquerque Journal*, February 27, 1927; "Robbers Caught at Jemez Springs, Sheriff, with Deputies Lies in Wait with Shotguns Ready," *Albuquerque Journal*, March 18, 1931.
9. "Two Dead in Midnight Duel," *Santa Fe New Mexican*, November 25, 1912. The *New Mexican* reported on December 2, "Nathan Salmon returned last night from Jemez Pueblo, and reports that his former partner, Moses Abousleman, who was wounded trying to round up some alleged sheep thieves, is resting easy but can not be moved for a week to ten days." On May 27, 1913, the *New Mexican* reported, "Nathan Salmon is just in receipt of a letter from M. Abousleman, his former business partner who was shot and dangerously wounded about three months ago, and who has steadfastly refused to allow his arm to be amputated, although told it would mean his death."
10. *Santa Fe New Mexican*, May 27, 1913.
11. *Albuquerque Journal*, March 18, 1931.

CHAPTER 10

1. Niederman, *Quilt of Words*, 53–54.
2. "Case Car Makes a Trip to Jemez and Return," *Albuquerque Morning Journal*, January 20, 1912.

CHAPTER 11

1. "Jemez Pueblo Keeps Title to Portion of National Preserve," Associated Press, May 23, 2023.
2. Bowden, "Canon de San Diego Grant."
3. Ebright, *Land Grants and Lawsuits in Northern New Mexico*, 151.
4. "Overview of New Mexico Politics, 1848–1898," History, Art & Archives, United States House of Representatives, https://history.house.gov/Exhibitions-and-Publications/HAIC/Historical-Essays/Continental-Expansion/New-Mexican-Politics, accessed July 16, 2024; Ebright, *Land Grants and Lawsuits in Northern New Mexico*, 42–43.
5. Ebright.
6. "Canyon of San Diego Land Grant," Jemez Valley History, https://Jemezvalleyhistory.org/?page_id=4514, accessed July 16, 2024. This undated partial document was probably written by John Amos Adams Sr., because he was an early surveyor for the Forest Service in 1912 in the Jemez and in that job he would likely have become aware of the details of the land grant partition and the corrupt sale described in the document. The website includes a hand-drawn map, from around 1906, of land grantee heirs (and subsequent owners) of parcels along the Jemez River, surrounded by the community land grant. Unfortunately, it is a low-resolution scan and many details are illegible. I have been unable to locate the original.
7. deBuys, "Fractions of Justice."
8. Glover, *Jemez Mountains Railroads*.

Notes

CHAPTER 12

1. On the history of Montezuma Castle (Hotel), see James H. Purdy, Montezuma Hotel Complex registration form, National Register of Historic Places Inventory, National Park Service, 1974, https://www.nps.gov/subjects/nationalregister/database-research.htm.

2. An excellent written and photographic history of the Santa Fe Northwestern Railroad is Glover, *Jemez Mountains Railroad*. In his history of the upper Jemez Valley (Scurlock, "Euro-American History of the Study Area"), Dan Scurlock includes this paragraph about the Oteros and the aborted plan to build a passenger line to Jemez Springs:

> Probably the first successful businessmen in the area were Mariano S. Otero and his uncle Miguel Antonio Otero. In 1881 the Oteros jointly owned the Jemez Springs, and Mariano owned the Sulphur Springs, located approximately two miles northwest of Redondo Creek and just outside (according to the 1876 survey) the west boundary of the Baca. To operate these resorts a development company was formed with Miguel Otero as president and Mariano Otero and various officials of the Atchison, Topeka, and Santa Fe Railroad on the board of Directors. Bath houses, a hotel, and other developments were initiated.... The surveying and grading of one mile of a branch railroad which would bring tourists From Bernalillo co Jemez Springs were completed in 1881, and the hotel and bath houses at Jemez Springs were completed in 1882. However, the death of Miguel Otero later in the year stopped the development of the spur railroad, and plans for making Jemez Springs into a major resort were dropped.

3. As noted in chapter 11, John Adams stated, "The White Pine Lumber company bought the grant in 1910 for a price reported to have been more than $400,000. This company, which became the New Mexico Lumber and Timber company in 1920, held the grant as a reserve supply of timber until 1922." That history is not consistent with Glover's explanation. Glover says WPL bought the timber rights from the Jemez Land Company in 1922 for more than $800,000 based on the estimated board feet of the grant. WPL was eventually bought out by NML&T. If Glover is correct, and I suspect he is, the

profit that lawyer McMillen, banker Raynolds, and other owners of the Jemez Land Company reaped from their Cañon de San Diego Grant swindle of the original land grant heirs may have been more than twice what I estimated in chapter 11.

4. Sando, *Nee Hemish*, 93–103, provides a detailed history of the Santa Fe Northwestern Railroad, including the legal wrangling over right-of-way through Jemez Pueblo lands, the Pueblo Lands Condemnation Act and its effects, and subsequent congressional actions.
5. Sando, *Nee Hemish*, 94.
6. It appears that the Precinct 10 census counted workers and families living in the Porter and Gilman areas, although technically those settlements were in Precinct 9, Cañon de San Diego. Almost all the new family heads in 1930 are listed as railroad men or logging and lumber camp workers. The men and families were from all over the United States and Canada, with most coming from southeastern states.
7. Sando, *Nee Hemish*, 98.

CHAPTER 13

1. Baker et al., *Timeless Heritage*, especially notes 3 and 6; "5450 NFS Modification, Santa Fe, Jemez National Forest," Southwestern Region, US Forest Service, Albuquerque. This file, containing 351 pages of reports, memoranda, and maps, provides many historical details of the Jemez National Forest, including the 1904 Benedict and Reynolds report.
2. Baker et al., *Timeless Heritage*.
3. Tucker and Fitzpatrick, *Men Who Matched the Mountains*.
4. Davis and Slonaker, "First Forest Service Wireless." The first page of this article includes a biography of John A. Adams Sr., including a history of his career with the Forest Service. In addition to surveying and other jobs, he was involved in early experiments with the use of radios by the Forest Service.
5. Tucker, *Early Days*, 92–97.
6. Tucker.
7. Transcriptions of daily diaries of early southwestern rangers on file at Southwestern Region, US Forest Service, Albuquerque.
8. "Ranger Killed in Auto Crash," *Albuquerque Tribune*, November 2, 1938

Notes

9. Tucker and Fitzpatrick, *Men Who Matched the Mountains*, 240–43, includes a description of Fred Swetnam's early career and Grace's role in helping manage the districts they were on. This includes an amusing story: Grace was in labor with their second child (Michael), and they were on the way to the hospital. They saw some escaped horses, so they stopped. She held the gate, notwithstanding labor pains, while Fred rounded up the horses into the corral.

CHAPTER 14

1. Swetnam, "Peeled Ponderosa Pine Trees." This paper includes a review of historical mentions of peeled trees in the western United States since 1805 and tree-ring dating of some of these trees in New Mexico. Methods for sampling and tree-ring dating peel scars are also described and illustrated.
2. Benedict and Reynolds, *The Proposed Jemez Forest Reserve, New Mexico*. A mention of the use of peeled trees for food in the Pecos, New Mexico, area is in Johnson, *Reminiscences of a Forest Ranger*. Regarding the use of pine inner bark, a man who was part Pecos (Towa) and part Hispanic told him, "Also they would eat it fresh and raw if the need was great." However, Opler, *Myths and Tales of the Jicarilla Apache Indians*, quotes an informant who says, "About June the pine tree has bark ready for you. That is good food too."
3. See Swetnam, "Peeled Ponderosa Pines Trees," 181–84, on methods of sampling and cross-dating peeled trees.
4. Swetnam, "Peeled Ponderosa Pine Trees at the Warm Spring."
5. Swetnam, 177–87.
6. Swetnam, 187.
7. Swetnam, "Peeled Ponderosa Pine Trees at the Warm Springs." Other works on New Mexico peeled trees include Corral, "Bark Substance Utilization in the Warm Springs Area"; Kaye and Swetnam, "An Assessment of Fire, Climate, and Apache History in the Sacramento Mountains"; Maingi and Swetnam, *Peeled Ponderosa Pine Trees from Northern New Mexico*; Martorano and Beardsley, "New Mexico's Forgotten Resource"; Sheppard et al., "Dendrochronology of Culturally Modified Trees"; and Towner and Galassini, "Cambium-Peeled Trees in the Zuni Mountains, New Mexico."
8. Johnson, *Final Report on the Battle of Cieneguilla*; Gorenfeld, "The Battle of Cieneguilla."

9. According to Alice Bullock, author of *The Squaw Tree: Ghosts and Mysteries of New Mexico*, some foresters called peeled pines "squaw trees." My father used the term when he first showed me peeled ponderosa pines in the 1960s. The term *squaw* for an Indigenous woman is now a widely recognized derogatory and racial slur. Bullock apparently did not understand the ugly baggage of the name in 1978, but she made a useful inference about its origin. She quotes a forest ranger: "'It's the women,' older heads [of tribes] have told the rangers. 'When game was scarce, they had trouble taking care of the children they had, and they didn't want any more. A woman with child would chew the gum the sap made, and she would have no child.'"

Bullock goes on to say that it is well-known among ranchers that if cows eat the green needles of ponderosa pines, they will abort. She concludes her chapter with these statements: "Often old wives tales carry more than a kernel of hard-earned wisdom. Did Indian women, either by observing animals or by sheer accident, learn that something in the yellow pine would cause a miscarriage? We're not sure on that score. We do know that scarred old pine trees, somewhere along the way, picked up the beguiling and perplexing name, 'squaw tree.'"

In her 1996 master's thesis, Patricia Corral addressed the subject of inner bark possibly used as an abortifacient. Here is my summary of this topic from Swetnam, "Peeled Ponderosa Pine Trees at the Warm Springs," drawing from Corral's work and from veterinary literature on the effects of pine tree compounds on cattle and mice:

> Corral . . . interviewed one Hispanic man [in his fifties] who had grown up near Vadito who stated he had heard as a young man that a "medicinal tea" was made from pine inner bark that "was used by women to abort their fetuses when they were sick or when there was a particularly harsh winter." Corral also cited a personal communication with Camino Real District archaeologist Robert Kreibel who had been told by an elderly Picuris woman: "We used that to bring on our periods." [In direct interviews by Corral with seven Picuris women, six of them said they had no knowledge of bark peeling except for use of outer bark for firing pots. A seventh elderly Picuris woman said she heard from her grandfather about using inner bark as an occasional food, but no mention was made of medicinal uses.] In the same discussion, Corral lists a Carson National

Forest "Heritage Resources Report" by Feiler ... wherein it is noted that "cattle are known to miscarry after ingesting pine needles."

In my own literature searches on this topic, I found more than a dozen veterinary medicine articles supporting the fact that pine needle consumption, and specifically the compound isocupressic acid in pine tissues, causes abortion in cattle. Laboratory studies also have shown that pine needle ingestion causes abortion in mice. It is unknown at this time [from controlled pharmacological studies] if pine inner bark contains compounds that might act as an abortifacient in humans. Perhaps future studies could investigate this question. For now, it seems that this possible use is consistent with an overall hypothesis that pine inner bark was used as food or medicine during times of unusual duress.

CHAPTER 15

1. "Runner Attacked by Bear in Valles Caldera," *Santa Fe New Mexican*, June 25, 2016.
2. Holmes, *Random Records of a Lifetime Devoted to Science and Art*, 259–75. Incidentally, Holmes was very familiar with grizzly bears, having served on previous US Geological Survey expeditions in Yellowstone and elsewhere in the northern Rockies.
3. "Returned from Jemez: A Bear Story in Which Messrs Finical and Powars Figure," *Albuquerque Daily Citizen*, June 24, 1898.
4. Isaacs, *Jemez Mountains 100 Years Ago*, 49–51.
5. Brown, *Grizzly in the Southwest*.

CHAPTER 16

1. Several excellent sources for this history include Sando, *Nee Hemish*; Liebmann, *Revolt*; and Elliott, National Historic Landmark nomination form.
2. Bowden, "Canon de San Diego Grant."
3. Cañoncito is labeled on the 1876 Wheeler Survey map; see chapter 1.

4. There are multiple accounts of Navajo, Jicarilla Apache, Ute, and Comanche raiding and fighting with Spanish, Pueblos, and then Americans prior to the 1860s in Northern New Mexico and in the Jemez Mountains specifically. The raiding and fighting was especially frequent and intense during the early to mid-1800s, with entire herds of sheep stolen and numerous Pueblo people, colonists, and travelers killed. Accounts of Navajo battles with Hemish, and with American soldiers in the Jemez Mountains during the early 1800s, are included in Sando, *Nee Hemish*, 12, 134–37; Scurlock, *Euro-American History of the Study Area*; Keleher, *Turmoil in New Mexico*; and Anschuetz and Merlan, *More Than a Scenic Mountain Landscape*.
5. Huning, *Trader on the Santa Fe Trail*.
6. A caption handwritten on a photograph taken by Townshend on his 1903 visit to Jemez Springs provides a clue about the timing of first settlement by San Diego land grantees in the Jemez Springs area. The photo shows the entrance to a cave in what appears to be volcanic tuff, with the ruins of a built-in stone doorway with a wooden lintel. The caption reads, "Cave once inhabited by Francisco Archuleta, the first man to plant a crop[?] at the Jemez Hot Springs." José Francisco Archuleta (1807–1880) and his son Emiterio (1844–?) were friends of Townshend's when he briefly lived in the Jemez Valley in the 1870s. He visited Emiterio in 1903 when he took this photo. Undoubtedly, either Francisco or Emiterio told Townshend that Francisco was the first to farm at Jemez Hot Springs. Francisco would have been about fifty when Franz Huning took the baths at his log cabin bathhouse in 1857. The killing of José Francisco and Maria's two sons by Navajo, as related in Townshend's stories (see chapter 19), attests to the fact that Jemez Hot Springs and surrounding areas were dangerous for settlement before the defeat of the Navajo by Kit Carson and the US Army, and the beginning of the Long Walk in spring 1864.
7. Townshend, *A Tenderfoot in Colorado*.
8. Townshend, *A Tenderfoot in New Mexico*.
9. Townshend, 25.
10. Townshend, 26.
11. "Francisco Perea, 1830–1913," US Government Publishing Office, https://www.govinfo.gov/content/pkg/GPO-CDOC-108hdoc225/pdf/GPO-CDOC-108hdoc225-2-2-4.pdf, accessed July 24, 2024.
12. Townshend, *Last Memories of a Tenderfoot*.

Notes

CHAPTER 17

1. For a scholarly and detailed description of Penitente history, see Chavez, "The Penitentes of New Mexico."
2. Townshend, *Tenderfoot in New Mexico*, 48–53.
3. Chavez, "Penitentes."
4. Townshend, *Tenderfoot in New Mexico*, 48.
5. Chavez, "Penitentes."

CHAPTER 18

1. LeCompte, "The Hispanic Influence on the History of Rodeo."
2. LeCompte.
3. Townshend, *Last Memories of a Tenderfoot*.
4. There is a stunning allegorical scene involving Correr El Gallo at Jemez Pueblo in N. Scott Momaday's Pulitzer Prize–winning novel *House Made of Dawn*. The Hemish protagonist of the novel, named Abel, is a PTSD-damaged WWII veteran. He has returned to Walatowa after the war, and one of the first village events he partakes in is the Correr. During the tug-of-war, a large "white man" seizes the rooster's corpse and proceeds to beat Abel viciously with the bloody prize. Then we learn that the white man is an albino Hemish.

CHAPTER 19

1. Sando, *Nee Hemish*, 107; Elliott, National Historic Landmark nomination form.
2. David Whitely, former pastor of Jemez Springs Community Presbyterian Church, kindly loaned to me several bankers boxes filled with historical documents, publications, and photographs pertaining to church history. These include records of establishment of the church, Shields family history, and the land gift from Francisco Archuleta, plus newsletters, centennial celebration (1981) documents, and much more. Documents also provide names of pastors, church members, and elders over the decades. I have scanned most of the key documents into pdf files, and these provide a basis for my historical

interpretations relative to this church in Jemez Springs. Frederick R. Sanchez's "The Presbyterians in the Valley of the Holy Eucharist, the Presbyterian Missionary Movement in the Late 19th and Early 20th Centuries in Jemez, New Mexico" summarizes Presbyterian church history in New Mexico and Jemez as well as Mary Stright Miller's life in the valley.

3. Cather, *Death Comes for the Archbishop*.
4. Horgan, *Lamy of Santa Fe*; Webster. "José Martínez and the New Mexico Reformation."
5. Brackenridge and García-Treto, *Iglesia Presbiteriana*.
6. Jemez Springs Community Presbyterian Church records and one hundredth anniversary celebration booklet.
7. Jemez Springs Community Presbyterian Church records; José Inez Perea biography, Menaul Historical Library of the Southwest, Albuquerque.
8. Anschuetz and Merlan, *More Than a Scenic Mountain Landscape*, 28.
9. Jemez Springs Community Presbyterian Church records.
10. Townshend, *Last Memories of a Tenderfoot*.

CHAPTER 20

1. Townshend, *Last Memories of a Tenderfoot*, 235–37.
2. Townshend, 123–36.
3. This mention of Juan Sandoval living in a cave temporarily somewhere near the Rio Cebolla, where the Sandovals had a homestead, is interesting. In note 6 in chapter 16 I refer to a photograph taken in 1903 by Townshend, showing a cave, with the young Hugh Miller and Maclean Fenton in the opening, and the caption, "Cave once inhabited by Francisco Archuleta, the first man to plant a crop at the Jemez Hot Springs." This photo was clearly taken by Townshend along the Rio Cebolla in 1903. I suspect the caption, possibly written by Townshend, is incorrect and should say that the cave was once inhabited by Juan Sandoval (not Francisco Archuleta, Sandoval's father in-law). Despite this probable error, the statement that Archuleta was the first man to "plant a crop at Jemez Hot Springs" still stands as credible.
4. According to the *Albuquerque Morning News* of January 12, 1919, "A.P. Vanderveer died at his home on the Cebolla last Monday of pneumonia, following a short illness of influenza. His death was a real shock to the community.

Notes

E. M. Fenton, father of Mrs. Vanderveer, was summoned from Albuquerque, and will conduct funeral services. Mr. and Mrs. Vanderveer entertained their neighbors from several miles around at a Christmas dinner and tree, and they and several of the guests were stricken with influenza within a few days. Mrs. Vandeveer and her 2 year old daughter [Jean] are still ill, though not seriously so. Also the several Fluke families."

5. Fitzgerald Family Tree, Jean Lime (Vanderveer) Fenton, Ancestry.com, accessed July 19, 2023.

CHAPTER 21

1. Townshend, *Last Memories of a Tenderfoot*, 123–36.
2. Swetnam, "Late Nineteenth and Early Twentieth Century Logging in the Jemez Valley."

CHAPTER 22

1. Very informative papers describe the origin and structure of fissure ridge travertines, especially in Turkey and Iran. Good examples, with excellent diagrams and photos, include Brogi et al., "Evolution of a Fault-Controlled Fissure-Ridge Type Travertine Deposit in the Western Anatolia Extensional Province" and Mohajjel, "Quaternary Travertine Ridges in the Lake Urmia Area."
2. The classic geological description and dating of Soda Dam were carried out by Fraser Goff of Los Alamos National Laboratory and colleagues. They also laid out the theory of lake stands in Valles Caldera and its association with major travertine dam formation episodes. Goff and Lisa Shevenell, "Travertine Deposits of Soda Dam"; Goff et al., "The Valles Caldera Hydrothermal System"; Goff and Goff, *Overview of the Valles Caldera (Baca) Geothermal System*. Also see Tafoya, "Uranium-Series Geochronology and Stable Isotope Analysis of Travertine from Soda Dam," which provides more dating evidence of the multiple deposits at Soda Dam.
3. Alexander, "The Excavation of Jemez Cave," provides a preliminary report on the Jemez Cave excavation. The paper includes the statement about workmen

camping in caves beneath Soda Dam during the winter of 1934 to 1935 (see chapter 25). Many more details came with the full published report: Alexander and Reiter, *Report on the Excavation of Jemez Cave, New Mexico*. Richard Ford reanalyzed faunal and botanical materials from Jemez Cave in the 1970s; see Ford, *The Cultural Ecology of Jemez Cave*.

4. Comparisons of old photos of Soda Dam in sequence, from older to younger, clearly show travertine growth of the cascade domes over the river. For example, a particular log shows up on the far east end in a 1903 photo by Townshend. This log is visible leaning up against the dome over the river but is not covered by travertine. By 1912, in photos by Jesse Nusbaum, the log is partially encased by travertine. In 2015 we again located this log, completely covered beneath the dome on the east side, just above a small "window" overlooking the waterfall. About twenty inches of travertine had accumulated over the log in about 110 years.

CHAPTER 23

1. Goff and Shevenell, "Travertine Deposits of Soda Dam"; Tafoya, "Uranium-Series Geochronology and Stable Isotope Analysis of Travertine from Soda Dam."
2. Goff and Shevenell, "Travertine Deposits of Soda Dam."
3. Goff et al., "The Hydrothermal Outflow Plume of Valles Caldera."
4. Goff and Shevenell, "Travertine Deposits of Soda Dam."
5. Depth to the crystalized top of the magma chamber is five to seven kilometers (3.1 to 4.3 miles) beneath the southwest portion of the Valles Caldera, as detected by various seismic studies, most recently: Wilgus et al., "Shear Velocity Evidence of Upper Crustal Magma Storage beneath Valles Caldera."
6. The three youngest rhyolite eruptions are now formerly named the Banco Bonito Flow, the Battleship Rock Ignimbrite, and the El Cajete Pyroclastic Beds. The new geologic map of Valles Caldera shows a discontinuous gravel deposit beneath Banco Bonito and above El Cajete, discovered during the remapping project (Goff et al., *Geological Map of the Valles Caldera*). The new map also shows that El Cajete and Battleship Rock are co-magmatic. The ignimbrite is interlayered in the pumice beds, and they both originate from El Cajete crater. Goff et al., *Geological Map of the Valles Caldera*.

Notes

CHAPTER 24

1. I formally requested documents pertaining to the dynamiting of Soda Dam and maintenance of the highway through the dam from the New Mexico Department of Transportation using the New Mexico Inspection of Public Records Act, 14-2-1 NMSA 1978. I received scanned copies of the engineers' designs and surveying records for the roadway, including the road leveling through Soda Dam and construction of the French drain beneath the roadway. The plans show handwritten, dated notations indicating they are final "as built" designs. They indicate that planning and then construction was sometime between June 1960 and September 1961, but the exact date of the blasting was not included. NM DOT suggested I look further at the chief engineer's annual reports. With kind assistance from Laura Calderone of New Mexico State Library, I received copies of the early 1960s reports. These reports indicate that the contract for the work was issued in October 1960 and the job was completed in July 1961. So it is possible the blasting was done in late 1960 or early 1961. I suspect that early 1961 is most likely, so this is the date I use (for now). Finally, NM DOT also sent a spreadsheet of maintenance activities for NM 4 during the 2000s, but they were irrelevant to Soda Dam, never mentioning it. I suspect someone in that office just grabbed a file to try to appease my formal request rather than finding any actual, relevant data on Soda Dam roadway maintenance.
2. "4500 Acre Tract of San Diego Land Grant Near Jemez to Be Developed in Homesites," *Albuquerque Journal*, July 8, 1962; full-page ad in the *Albuquerque Journal*, August 19, 1962, with map and text reading, "Only 60 miles north of Albuquerque on paved Highway 4, Choice Sites Now Available [in] Jemez Country."
3. Early descriptions of Soda Dam refer to many springs around and on the dam. An example is from Wheeler, *Explorations and Surveys West of the One Hundredth Meridian*:

> The upper group consists of forty-two springs; the taste of the water is somewhat like Vichy; the temperature ranges between 70 deg F and 105 deg F., with but few exceptions; the surface of each is less than 1 square foot. The most of these springs originate in cones and mounds, consisting of spring-deposit, chiefly carbonate of lime. One of these mounds is 20

feet high, 10 feet wide and 130 feet long, with twenty-two springs on it. A few yards above, and at right angles to it, is another mound, 30 feet high, about 200 feet long, and 15 feet wide.

And see J. K. Livingston's description in his 1886 article, quoted in chapter 7. In his recollections extending back to the 1920s, Joseph E. Routledge said this about Soda Dam: "There were springs above the Soda Dam that had really formed a dam across the Jemez River and canyon during hundreds of years of soda deposits. We think there was a great flood, and the river washed out around the right side of the dam and now appears to flow through it. In our day, there were two open pits 50 yards above the dam, where you could bathe alfresco in the tepid water. There were signs of old bath houses there" (Isaacs, *Jemez Mountains 100 Years Ago*, 12–13). Partly confirming Routledge's description of soaking pools on the north side, a 1960 NM DOT survey map of the road construction through Soda Dam shows this notation about where the current parking lot is located on the west side of the road above the dam: "14' x 8' spring."

4. Soda Dam blasting documents, New Mexico Department of Transportation, 1960–1961.
5. "Wildlife Group Works to Preserve Soda Dam," *Albuquerque Tribune*, April 12, 1969.
6. J. W. Little, "Soda Dam to Be Preserved through Efforts of Wildlife Group," *Outdoor Reporter*, May 1978, 6–7.
7. "Forestry Takes Over Soda Dam," *Los Alamos Monitor*, March 31, 1978.
8. Adler and Stokely, *Jemez Cave Condition Assessment*.

CHAPTER 25

1. J. K. Livingston, "Visit to the Valles, Jemez Springs and Monument Cañon, a Veritable Land of Modern Wonders," *Santa Fe New Mexican*, July 18, 1886.
2. Alexander, *Excavation of Jemez Cave*.
3. Alexander's statement that the caves beneath Soda Dam provided the "ideal camping" site during the winter for "4 men," along with Livingston's statement about "a certain grotto on the north side of the dam," are the most direct eyewitness evidence discovered to date of a large cave (or caves) on the

north side. Nevertheless, some uncertainty and ambiguity remain because the descriptions of a large cave by Livingston and others seem to at least partly describe the grotto on the southeastern end of Soda Dam near the waterfall. Interestingly, a 1903 photo by R. B. Townshend shows that the southeastern grotto at one time did have a window overlooking the waterfall, as Livingston mentioned. That window has now been closed by travertine flow over the dome above. The very wet and cramped spaces of the southeastern grotto today would certainly not serve as an ideal campsite or as a place for a "library" or "reception room." Perhaps new technology, such as ground penetrating radar, may someday confirm or refute the existence of a large cave or caves under the northwestern end of Soda Dam.

4. Pettitt, *Exploring the Jemez Country*, 16.

CHAPTER 26

1. The latest and perhaps most comprehensive assessment of radiometric ages of Valles Caldera volcanism is in Nasholds and Zimmerer, "High-Precision 40Ar/39Ar Geochronology and Volumetric Investigation of Volcanism and Resurgence Following Eruption of the Tshirege Member" and Zimmerer et al., "Eruptive and Magmatic History of the Youngest Pulse of Volcanism at the Valles Caldera."
2. Goff and Shevenell, "Travertine Deposits of Soda Dam."
3. The lidar data used in this map was acquired in late June and early July 2012 for the US Forest Service's Southwest Jemez Collaborative Forest Landscape Restoration Program for the southwestern portion of the Santa Fe National Forest. The lidar survey flight was made by Watershed Sciences of Portland, Oregon.
4. My son Tyson L. Swetnam and colleagues have used lidar in forest ecology, including in the Jemez Mountains. See Swetnam et al., "Discriminating Disturbance from Natural Variation with LiDAR in Semi-arid Forests in the Southwestern USA." The point cloud created by a lidar scan includes all objects the laser intercepts, including trees, shrubs, and the ground surface. The ground surface layer (map), as shown in this chapter, is derived by subtracting all objects above the ground level using computer algorithms. Likewise, the aboveground objects can be identified (segmented) as trees and so on, and those can then be analyzed for height, area, spacing, and other traits.

5. Fink and Manley, "Explosive Volcanic Activity Generated from Within Advancing Silicic Lava Flows."
6. Goff et al., *Geological Map of the Valles Caldera*.
7. Explosion craters of this type, where hot eruptions vaporize water and underlying deposits to cause steam blasts, are relatively common volcanic features, a type of phreatic explosion crater. Fink and Manley, "Explosive Volcanic Activity."
8. Roos et al., "Native American Fire Management at an Ancient Wildland–Urban Interface in the Southwest United States."
9. Roos et al.

CHAPTER 27

1. "Jemez River Near Jemez, NM—08324000," US Geological Survey, datahttps://waterdata.usgs.gov/monitoring-location/08324000/#parameterCode=00065&period=P7D, accessed July 29, 2023.

    In Sando, *Nee Hemish*, 78–77, the 1941 flood is listed with an estimated maximum flow of six thousand cubic feet per second on May 15 and the 1958 flood with a maximum of fifty-nine hundred cubic feet per second on April 21. (Note that the graph in this chapter shows average daily flows rather than maximum flows.) Sando also says the 1941 flood destroyed portions of the Santa Fe Northwestern Railroad lines, an irrigation diversion dam, parts of fields, and a flour mill at Jemez Pueblo.
2. "Ol' Rio Grande Betrays Wicked Side Every So Often; '74 Flood the Worst," *Albuquerque Journal*, May 27, 1948.
3. Scurlock, *From the Rio to the Sierra*, 32–38. Scurlock writes, "1941: (May) (early to late) The most severe flood since at least 1890 occurred on the Jemez River near Jemez Pueblo. . . . The Guadalupe and Jemez rivers flooded and washed out 3 miles of track of the Santa Fe Northwestern Railroad" (78).
4. "Two Men Have Narrow Escape from Drowning in Waters of Jemez," *Albuquerque Morning Journal*, April 16, 1915.
5. "Swirling Floodwaters Isolate Jemez Springs," *Albuquerque Journal*, April 22, 1958.
6. "Tribune Camera Pictures Jemez Area Flood," *Albuquerque Tribune*, April 25, 1958.

Notes

CHAPTER 28

1. Kues, "Guide to the Late Pennsylvanian Paleontology of the Upper Madera Formation."
2. Lucas et al., "Lithostratigraphy, Paleontology, Biostratigraphy."
3. "The Bisti Beast," New Mexico Museum of Natural History & Science, 2024, https://www.nmnaturalhistory.org/exhibits/permanent-exhibits/bisti-beast.
4. Rowe et al., "Human Occupation of the North American Colorado Plateau"; Harry Baker, "37,000-Year-Old Mammoth Butchering Site May Be Oldest Evidence of Humans in North America," *Live Science*, August 10, 2022, https://www.livescience.com/mammoth-butchering-site-new-mexico.
5. Morgan and Rinehart, "Late Pleistocene (Rancholabrean) Mammals from Fissure Deposits in the Jurassic Todilto Formation."
6. Heintzman et al., "A New Genus of Horse from Pleistocene North America."
7. Morgan and Lucas, "Pleistocene Vertebrate Faunas in New Mexico."
8. Morgan and Lucas, 192.
9. Taylor et al., "Early Dispersal of Domestic Horses into the Great Plains and Northern Rockies"; Curry, "Horse Nations."

CHAPTER 29

1. Swetnam, "Long-Term Vulnerability and Resilience of Coupled Human-Natural Ecosystems to Fire Regime and Climate Changes at an Ancient Wildland Urban Interface," National Science Foundation Award Abstract #1114898, 2011.
2. Roos and Swetnam, "Fire Adds Richness to the Land."
3. Liebmann et al., "Native American Depopulation, Reforestation, and Fire Regimes."
4. Swetnam et al., "Multi-Scale Perspectives of Fire, Climate, and Humans in Western North America and the Jemez Mountains." For another extensive tree-ring and fire scar study in the Jemez Mountains, see Dewar et al., "Valleys of Fire."

5. Liebmann et al., "Native American Depopulation, Reforestation, and Fire Regimes," figure 2; Farella and Swetnam, "Terminus Ante Quem Dating."
6. Roos et al., "Indigenous Fire Management."
7. Roos et al., "Fire Suppression Impacts."
8. Abatzoglou and Williams, "Impact of Anthropogenic Climate Change."

CHAPTER 30

1. "Jemez Blaze Threatens Summer Homes," *Albuquerque Journal*, June 6, 1971.
2. Westerling et al., "Warming and Earlier Spring Increase Western US Forest Wildfire Activity," a paper I coauthored in 2006, showed that earlier and warmer springs had led to increased numbers of large wildfires in the western United States since about 1980. Since this paper was published, many other scientific papers have confirmed these trends and the linkages between greenhouse gas–caused global and regional warming and wildfires. See, for example, Williams et al., "Temperature as a Potent Driver of Regional Forest Drought Stress and Tree Mortality."
3. "Help Came Quickly in Jemez Fire," *Albuquerque Journal*, June 12, 1971.
4. "Willing Work by Volunteers Aided at Fire, but Safety First," *Albuquerque Journal*, June 13, 1971.
5. "A Visit to a Tragic Scene—Where a Forest Died," *Albuquerque Tribune*, June 17, 1971.
6. "The 'Wild Beast' That Devours the Dry Forests," *Life*, July 23, 1971, 28–30.
7. Forthofer, "Fire Tornadoes."
8. "Las Conchas Fire," InciWeb, 2011, https://web.archive.org/web/20111016012328/http://inciweb.nwcg.gov/incident/2385.
9. John Schwartz, "As Fires Grow, a New Landscape Appears in the West," *New York Times*, September 21, 2015. This article includes photos of the Las Conchas Fire burned area and quotes from interviews with me and Craig Allen.
10. On type conversion of forests to shrublands and grasslands in the Southwest, see Guiterman et al., "Vegetation Type Conversion in the US Southwest." On type conversion in the Jemez Mountains, see Guiterman et al., "Dendroecological Methods for Reconstructing High-Severity Fire in Pine-Oak Forests"; Guiterman et al., "Long-Term Persistence and Fire Resilience"; Roos and Guiterman, "Dating the Origins of Persistent Oak Shrubfields."

Notes

11. The idea of using non-native species in restoration is controversial—understandably so given the uncertainties and potential unintended consequences. Still, as an emergency or last resort option, it might involve planting more drought-, heat-, and fire-tolerant pines from northern Mexico in parts of the US Southwest. On the general issue, see Schlaepfer et al., "The Potential Conservation Value of Non-Native Species." On the difficulties of restoring forests in the Jemez and in New Mexico, see O'Donnell, "What Does Forest Restoration in the US Southwest Look Like in the Age of Climate Change?," *Ensia*, October 19, 2022.
12. "New Mexico Developing Reforestation Center to Aid in Wildfire Recovery," *Santa Fe New Mexican*, January 10, 2023.

CHAPTER 31

1. Sando, *Nee Hemish*.
2. Wheeler, "US Geographical Surveys West of the 100th Meridian."
3. In Tosa et al., "Movement Encased in Tradition and Stone," Jemez elder and former governor Paul Tosa explained the meanings this way:

> The Jemez refer to this sacred mountain as Sée Tôoky'aanu Tûukwâ (Place of the Eagle) or Wâavêmâ Ky'ôkwâ (Want for Nothing Peak) or simply Wâavêmâ.... Wâavêmâ, however, is more than a pretty place to the Jemez people. It is a living entity and the source of all life. Like a mother, Wâavêmâ gives life to Jemez newborns and sustains them on earth. The waters flowing off its flanks irrigate Jemez fields and nourish Jemez bodies. At the end of life, Jemez spirits return to Wâavêmâ after shedding their earthly skins. Wâavêmâ is equivalent in Judeo-Christian belief to the Holy Spirit, the Garden of Eden, and Heaven all rolled into one: the domain where life began, from which life continues to spring forth, and to which the deceased return.

He also writes, "The term Wâavêmâ refers both to Redondo Peak specifically and the Valles Caldera region generally."

4. "Cerro Pelado Fire Was Started by US Forest Service Pile Burns, Agency Confirms More Than a Year Later," *Albuquerque Journal*, July 23, 2023.

5. "Calf Canyon/Hermits Peak Fire," InciWeb, 2022, https://web.archive.org/web/20220521021140/https://inciweb.nwcg.gov/incident/article/8049/68692.
6. Allen, "Montane Grasslands in the Landscape of the Jemez Mountains."
7. Swetnam et al. "Applied Historical Ecology." On tree invasion of montane grasslands in the Jemez Mountains, see Coop and Givnish, "Spatial and Temporal Patterns of Recent Forest Encroachment."
8. Investigation report on Cerro Grande prescribed fire, National Interagency Fire Center, May 18, 2000.
9. Hagmann et al., "Evidence for Widespread Changes in Structure, Composition, and Fire Regimes"; Pritchard et al., "Adapting Western North American Forests to Climate Change and Wildfires."

CHAPTER 32

1. Matthew Hurteau, Tom Swetnam, and Craig Allen, "Use All Tools—Including Fire—to Restore Forest Resiliency" *Santa Fe New Mexican*, April 23, 2022.

CHAPTER 33

1. Several papers show that the drought of 2000 to 2021 was the worst in about twelve hundred years and that about half of the magnitude of this drought was attributable to warming caused by anthropogenic greenhouse gases. See Williams et al., "Large Contribution from Anthropogenic Warming"; Williams et al., "Rapid Intensification," chapter 34.
2. In addition to the three photos shown in this chapter, the Library of Congress online archive includes two more in this series, showing the Mission de San José and Soda Dam, and those are numbered 22 and 26, respectively. Given this sequence of numbers and those missing, there may be additional Cook, Bass, and Robinson photos of the Jemez still to be located.
3. There is extensive literature on the ecological and hydrological dynamics of cottonwood bosques. An example from Arizona but with some relevance to the Fremont cottonwoods and introduced tamarisk species found along rivers in

Notes

the Jemez is Stromberg, "Dynamics of Fremont Cottonwood (*Populus fremontii*) and Saltcedar (*Tamarix chinensis*) Populations."

4. Many of the sheep that were grazed in the Valles Caldera or on the San Diego Land Grant during the summer were herded up and down San Diego Canyon at the beginning and end of the grazing season. This passage from Scurlock, "Euro-American History of the Study Area," provides numbers of livestock on the Baca Location in 1918: "Luciano Garcia and Clyde Smith, residents of Jemez Springs, began work for Otero as vaqueros during this period.... One of Smith's duties was counting and collecting fees on sheep and cattle as they were driven onto the Baca in late April or May. Otero charged 25 cents per sheep and one dollar per cow for summer grazing rights. An estimated 200,000 sheep and several thousand cattle were grazed on the prolific grasses of the Baca's valles at this time."

5. Leopold, *A Sand County Almanac*.

CHAPTER 34

1. On attribution of wildfire area increases in the western United States to greenhouse gas–induced warming, see Abatzoglou and Williams 2016, "Impact of Anthropogenic Climate Change." On increased area burned in the western United States and its association with warming and earlier springs, see Westerling et al., "Warming and Earlier Springs." On climate-fire associations in the western United States and Canada, see Kitzberger et al., "Direct and Indirect Climate Controls."
2. Williams et al., "Temperature as a Potent Driver."
3. Williams et al., "Temperature as a Potent Driver." See also Williams et al., "Causes and Implications of Extreme Atmospheric Moisture Demand."
4. Aiuvalasit and Jorgeson, "Combining Paleohydrology and Least-Cost Analyses."
5. Swetnam and Betancourt, "Mesoscale Disturbance."
6. Williams et al., "Large Contribution from Anthropogenic Warming"; Williams et al., "Rapid Intensification."
7. Swetnam and Betancourt, "Fire-Southern Oscillation Relations"; Kitzberger et al., "Contingent Pacific-Atlantic Ocean Influence."

8. Kitzberger et al., "Contingent Pacific-Atlantic Ocean Influence."
9. Williams et al., "Temperature as a Potent Driver."
10. Mann, *Our Fragile Moment*.

CHAPTER 35

1. On the history of Montezuma Castle (Hotel), see James H. Purdy, Montezuma Hotel Complex registration form, National Register of Historic Places Inventory, National Park Service, 1974, https://www.nps.gov/subjects/nationalregister/database-research.htm.
2. History of the Oteros, Baca Location No. 1, and the Stone Hotel in Martin, *Valle Grande*, 34–46.
3. "Ho for Jemez Hot Springs!," advertisement in *Las Vegas Daily Optic*, November 16, 1881.
4. "Jemez Hot Springs," *Las Vegas Gazette*, December 11, 1880.
5. Swetnam, "Tree-Ring Dates of the Stone Hotel."
6. Martin, *Valle Grande*, 42.
7. Martin, 42–44. See also Anschuetz and Merlan, *More Than a Scenic Mountain Landscape*, 125–28.
8. *Albuquerque Morning Journal*, May 22, 1903. See the photos in chapter 24 of six horses pulling and three men driving a freight wagon with mining equipment over the west end of Soda Dam in late summer 1903. The load looks like the top part of a vertical boiler, so maybe it was a replacement part for the boiler that exploded at Sulphur Springs in May 1903, as described in the *Albuquerque Morning Journal*.
9. Martin, *Valle Grande*; Anschuetz and Merlan, *More Than a Scenic Mountain Landscape*.
10. Author's conversation with John Merhege about Union Oil drillers and noxious gas incident, August 11, 2022.
11. "Valles Caldera National Preserve Acquires Property with Unique Volcanic Features," National Park Service, https://www.nps.gov/articles/valles-caldera-national-preserve-acquires-property-with-unique-volcanic-features.htm, accessed September 5, 2024.

Notes

CHAPTER 36

1. Rogers et al., first-day road log.
2. An intrepid German professor of geography, Dietrich Fliedner, presumably on sabbatical, spent a summer in about 1972 or 1973 hiking and climbing all over the slopes above and around Battleship Rock and down the canyon to the Unshagi village ruins. He mapped what he interpreted as culturally created features on the ground. These included dozens of field houses, rock shelters, trails, and "probable" agricultural fields. He mapped the Camp Shaver area as likely agricultural fields, with multiple ancient trails and field houses in the area. Fliedner, "Pre-Spanish Pueblos in New Mexico."
3. Scurlock, "Euro-American History of the Study Area," 138, 142.
4. Scurlock writes, "Reportedly Clyde Smith's father was removed from Battleship Rock due to this survey showing that his homestead was outside the grant on Forest Service land." Robert Cart, another descendant of James Smith, said the story he heard growing up was that Smith lost the property at Battleship Rock due to unpaid back taxes. The twenty-six acres the YMCA now owns is on the west side of the camp on what was once San Diego Land Grant property. I speculate that portion was donated to the Boy Scouts in the 1920s by Guy Porter of White Pine Lumber Company, which owned the grant at that time. The CCC and NYA buildings were later built on what is still national forest lands.
5. "Boy Scouts Build a Council Ring at Their Camp and Perform Colorful Ceremonies at Night," *Albuquerque Journal*, August 23, 1926. The article describes 150 Boy Scouts camping in twenty-five tents in the large clearing in the forest at Battleship Rock. The article also refers to Cathedral Canyon, Cathedral Falls, and Cathedral Camp, which were in close walking distance of the main Camp Porter, beneath Battleship. I suspect this area is the waterfall and slot canyon on the northwest side of Battleship Rock, beyond the present-day campground. During my seasons at Camp Shaver (1969–1972), we called this area Hidden Canyon and Hidden Falls.
6. "Townsfolk Put Out Fires in Forest Preserve" *Albuquerque Journal*, May 23, 1933. This article describes a wildfire at the camp at Battleship and says it was extinguished by various people from Jemez Springs, including Linus Shields: "They joined a number of men, women and children who wielded hats, sacks,

Notes

and every other available weapon in the fight against the blaze, which threatened for a time to destroy the big frame mess hall of the Boy Scout camp."

7. "Art and Archaeology in New Mexico, by Members of the Faculty of the University of New Mexico," *Albuquerque Journal*, February 16, 1930. According to this article,

> The site of the General Session is Battleship Rock Camp in the Jemez Canyon at the confluence of the San Diego River and the East Fork, 5 miles north of Jemez Hot Springs and 12 miles above the sole surviving village of the Jemez Group. The ruins of 18 or more towns lie in the valley and on the adjacent mesas. The program of the general session will be realized in courses of study with naturalists, ethnologists and archaeologists, supplemented by nature walks, sketching trips and archaeological examinations, excavations at the ruins of Un Shagi and Nona Shagi, wherein the students will assist by supervising the workmen, recovering, recording and preparing the material's uncovered for further study and museum installation; surveying, mapping and photographing; field trips to the pueblos of the Rio Grande Valley to observe their dramatic ceremonies, arts and industries.

There are two UNM reports on the excavations at Unshagi and Nanishagi, and at least one PhD dissertation. Along with Gíusewa, now at the Jemez Historic Site in Jemez Springs, these are among the few nearly completely excavated Hemish ancestral village sites.

8. An old photo from the US Forest Service labels the camp at Battleship "La Junta fly camp." Dirk Van Hart, in his book on CCC camps in New Mexico, says this camp was called Camp Kearny at one time (according to one source), supposedly after the American general who marched with troops into New Mexico in 1846 and seized the territory. Van Hart, *Camps and Campsites*, 45.

9. "U Buildings to Be Constructed," *Albuquerque Tribune*, April 25, 1938.

10. This organization is now known as The Arc of the United States, and it has state and local chapters all over the nation. The Arc was founded by parents of people with intellectual and developmental disabilities. The Arc advocates for disabled people, helps them find jobs, and helps employers adapt to their needs.

Notes

CHAPTER 37

1. Biographical information about the Rowland family was derived from family tree entries in Ancestry.com. Multiple mentions of Howard Rowland in the society pages of Albuquerque newspapers in the 1920s indicate that he was very active on the social scene. For example, the article "Cabot Recalls University Days Here; Goes to Ranch," *Albuquerque Morning Journal*, March 23, 1926, describes a visit by Bruce Cabot to Rowland's ranch in Jemez Springs.
2. "Cabot Recalls University Days."
3. This story was originally published in the *Jemez Thunder*, and it was recently republished in Isaacs, *Jemez Mountains 100 Years Ago*.
4. "Rowland Estate Value $199,336," *Albuquerque Tribune*, January 13, 1959; "Youth Camp to Open Soon," *Albuquerque Tribune*, May 27, 1959.
5. Articles and reports on the archaeology of Jemez Cave include Alexander, "The Excavation of Jemez Cave"; Alexander and Reiter, *Report on the Excavation of Jemez Cave*; Ford, "The Cultural Ecology of Jemez Cave"; Barbour, "The Early Farmers of Jemez Cave." Also see chapter 25.
6. Lipe et al., "Staying Warm in the Upland Southwest."
7. Alexander and Reiter, *Report on the Excavation of Jemez Cave*; Ford, "The Cultural Ecology of Jemez Cave"; Barbour, "The Early Farmers of Jemez Cave."

CHAPTER 38

1. "Father Gerald Fitzgerald," Servants of the Paraclete, 2001, https://web.archive.org/web/20090403163341/http://www.theservants.org/story.htm.
2. General articles about the SOP include "In New Mexico, Shadows of a Former Haven for Troubled Priests," *Santa Fe New Mexican*, May 30, 2022; "Early Alarm for Church on Abusers in the Clergy," *New York Times*, April 2, 2009.
3. The *New York Times* provides "a selection of letters to and from the Rev. Gerald M.C. Fitzgerald, who warned American bishops and even the Pope about sexually abusive priests as far back as 1952. The papers were unsealed by a judge in New Mexico in 2007 and authenticated in depositions with Father Fitzgerald's successors." See "Father Fitzgerald's Correspondence," *New York Times*, https://www.nytimes.com/interactive/projects/documents/

father-gerald-fitzgerald-correspondence-priest-sex-abuse, accessed August 6, 2024.

4. In Matthew 26:24, in response to a man who will betray him, Jesus says, "It would be better for him if he had not been born."

5. Letter from the archbishop of Santa Fe to Father Gerald Fitzgerald, August 23, 1965, https://www.bishop-accountability.org/docs/sP/SERVANTS_7367_7369__1965_08_23_Davis_to_Fitzgerald.pdf, accessed August 6, 2024. In this letter the archbishop issues eight orders to Fitzgerald, including instructions for turning over operational control of the SOP in Jemez Springs to other priest administrators (including Father Mooney) and for selling the island in the Caribbean that apparently Fitzgerald had purchased by this time. See also "Servants of the Paraclete," BishopAccountability.org, 2004, https://www.bishop-accountability.org/treatment/Servants/.

6. Others from the Catholic Church have pushed back on the conclusion that treatments at the SOP in Jemez Springs were ineffective. According to Stephen J. Rossetti in *Slayer of the Soul: Child Sexual Abuse and the Catholic Church* (Waterford, CT: Twenty Third Publications, 1991), from 1985 to 2008, 365 priests were treated for child sex abuse at the St. Luke Institute in Maryland and 22 of them, or 6 percent, relapsed. Even if a 6 percent relapse rate is accurate, which is questionable, clearly this should have barred any serial abuser priest from being returned to a parish, given that a couple of dozen (at least) such relapsed priests inflicted great harm to hundreds of children and their families. See also Thomas P. Doyle, "Paraclete Report," Richard Sipe: Priests, Celibacy, and Sexuality, http://www.awrsipe.com/doyle/2011/2011-01-11--paraclete_report.htm. This report describes an unpublished statistical study of pedophile recidivism following treatment at SOP. The study was pulled from a journal after submission at the request of SOP and was then locked in a safe.

7. According to "James Porter Dies; Ex-priest's Sex Crimes Foretold Scandal," Associated Press, 11, 2005, "Porter pleaded guilty in 1993 to molesting 28 children, but he once told a television reporter that he molested as many as 100 children during his time as a priest in the 1960s and early 1970s in the Fall River Diocese." See also "Camp Ped: Long after Roman Catholic Leaders Knew Pedo-Priests Couldn't Be Cured, Cardinal Roger Mahony Kept Packing Off His Worst Offenders to a Notorious New Mexico Rehab Center," *Los Angeles Times*, August 15, 2002. The article explains that Porter was a repeat veteran of the SOP in Jemez Springs, being sent there first for molesting children in 1967.

He was then released to a New Mexico parish. Amazingly, he went through two more cycles of treatment at SOP following additional child molestations and re-release to parishes. Another recounting of Porter's history of treatment and re-release by the SOP is in Bruni and Burkett, *A Gospel of Shame: Children, Sexual Abuse, and the Catholic Church* (New York: Perennial, 2002). This book mentions child molestations by Porter in Jemez Springs during the late 1960s as alleged by a "group of men" in 1992.

8. "Michael Stephen Baker," California Sex Abuse Lawyers, https://californiasexabuselawyers.com/catholic-priests/michael-stephen-baker, accessed August 6, 2024.

9. One of the most appalling accounts of the SOP history of recycling child molester priests well into the 1990s is "Camp Ped," *Los Angeles Times*, August 15, 2002 (see note 7). Despite the claim by an SOP administrator (quoted in "Camp Ped") that no additional priests accused of child molestation were treated at the SOP in Jemez Springs after 1995, Bruni and Burkett (see note 7) describe such a case sent to Jemez Springs in 1997.

10. *Sanctuary of Sin: How a Religious Order Became a Haven for Pedophile Priests*, NewsNation, 2023, https://www.newsnationnow.com/religion/sanctuary-of-sin-how-a-religious-order-became-a-haven-for-pedophile-priests.

11. For a description of the murder committed by Feit, his subsequent history, his conviction, and his death, see Erik La Garza, "Former Priest Convicted in Cold-Case Murder Dies in Prison," Courthouse News Service, February 13, 2020, https://www.courthousenews.com/former-priest-convicted-of-1960-murder-dies-in-prison.

12. The "Paraclete Report" (see note 6) is a lengthy and thorough review of SOP history, including details about how Feit became an administrator at the SOP in Jemez Springs before 1965, while the organization was still under Fitzgerald's leadership. In addition to Feit, the report lists three child-abusing priests who were sent to the SOP and then later appointed as staff members at SOP facilities, including one priest who committed child sexual abuse again after he left the SOP. He was convicted and sentenced to more than thirty years in prison. Further, the report states that Fitzgerald and Feit were allies in arguing (within the SOP and the archdiocese) that prayer and spiritual practices (rather than psychology and psychiatry) were the only ways to rehabilitate some problem priests.

13. I think the land for the park was donated by Fitzgerald to the village. Father Mooney (first name unknown) was appointed to an administrative job at SOP in 1965 (see note 5). I don't know why the boulevard was named after him.

CHAPTER 39

1. "Hippie Haven Double Slaying; Believe Borrowed Rifle Used"; "Police Hunt, Want to Quiz US Grant"; "Strange Life of Ulysses Grant—Portrait of Man Sought in Slayings," *Albuquerque Tribune*, December 2, 1970.
2. Ron Franscell, "Peace, Love . . . and Murder," Ronfrancell.com, September 3, 2020, https://ronfranscell.com/2020/09/03/hippies/; Harper Sullivan, "The Hippies of Placitas, Part II," My Strange New Mexico, December 8, 2020, https://mystrangenewmexico.com/2020/12/08/the-hippies-of-placitas-part-ii.
3. Email correspondence with Cathy Stephenson, January 28, 2023.
4. Eric McCrossen, "Jemez' Hot Spring Is Battlefield: Forest Service, Hippies May Clash Over Nudity," *Albuquerque Journal*, dated March 15, 1970.

CHAPTER 40

1. An 1876 Wheeler Survey map uses the name Sierra de los Valles.
2. Basso, *Wisdom Sits in Places*.
3. An example of place name stories and their significance to the Hemish people is Paul Tosa's description of the eagle and Wâavêmâ, now labeled Redondo Peak on most maps. See Tosa et al., "Movement Encased in Tradition and Stone."
4. This method has various names, including memory palace and method loci. See Frances A. Yates, *The Art of Memory* (London: Routledge and Kegan Paul, 1966).
5. Yates.

REFERENCES

Abatzoglou, John T., and A. Park Williams. "Impact of Anthropogenic Climate Change on Wildfire across Western US Forests." *Proceedings of the National Academy of Sciences* 113, no. 42 (2016): 11770–11.

Adams, John A. "Description and History, Cañon de San Diego Land Grant." Jemez Valley History, jemezvalleyhistory.org/?page_id=4514. Accessed August 9, 2024.

Adler, Rachel, and Sarah Stokely. *Jemez Cave Condition Assessment, Santa Fe National Forest, Jemez Ranger District, Sandoval County, New Mexico, March 2014*. Los Alamos: Bandelier National Monument, 2014.

Aiuvalasit, Michael J., and Ian A. Jorgeson. "Combining Paleohydrology and Least-Cost Analyses to Assess the Vulnerabilities of Ancestral Pueblo Communities to Water Insecurity in the Jemez Mountains, New Mexico." *American Antiquity* 89, no. 1 (2023): doi:10.1017/aaq.2023.67.

Alexander, Hubert G. "The Excavation of Jemez Cave: A Preliminary Report." *El Palacio* 38, nos. 18–20 (1935): 97–112.

Alexander, Hubert G., and Paul Reiter. *Report on the Excavation of Jemez Cave, New Mexico*. Albuquerque: University of New Mexico and School of American Research, 1935.

Allen, Craig D. "Montane Grasslands in the Landscape of the Jemez Mountains, New Mexico." Master's thesis, University of Wisconsin, 1984.

Anaya, Rudolfo, Francisco A. Lomelí, and Enrique R. Lamadrid. *Aztlán: Essays on the Chicano Homeland*. Albuquerque: University of New Mexico Press, 2017.

Anschuetz, Kurt F., and Thomas Merlan. *More Than a Scenic Mountain Landscape: Valles Caldera National Preserve Land Use History*. Fort Collins, CO: Rocky Mountain Research Station, US Forest Service, 2007.

Baker, Robert D., Robert S. Maxwell, Victor H. Treat, and Henry C. Dethloff. *Timeless Heritage: A History of the Forest Service in the Southwest*. College Station, TX: US Forest Service, 1988.

Bandelier, Adolf F. "The 'Montezuma' of the Pueblo Indians." *American Anthropologist* 5 (1892): 319–26.

Barbour, Matthew J. "The Early Farmers of Jemez Cave." *El Palacio*, Fall 2014.

Basso, Keith H. *Wisdom Sits in Places: Landscape and Language among the Western Apache*. Albuquerque: University of New Mexico Press, 1996.

Benedict, R. E., and R. V. Reynolds. *The Proposed Jemez Forest Reserve, New Mexico*. Albuquerque: Southwestern Region, US Forest Service, 1904.

Benavides, Alonso de. *Fray Alonso de Benavides Revised Memorial of 1634*, Vol. 4. Internet Archive. https://archive.org/stream/in.ernet. dli.2015.156868/2015.156868.Fray-Alonso-De-Benavides-Revised-Memorial-Of-1634-Volume-Iv_djvu.txt.

Bloom, Lansing B. "The West Jemez Culture Area." *New Mexico Historical Review* 21, no. 2 (1946): 120–26.

Bowden, J. J. "Canon de San Diego Grant." New Mexico State Record Center and Archives. https://newmexicohistory.org/2015/07/15/canon-de-san-diego-grant, 2014.

Brackenridge, R. Douglas, and Francisco O. García-Treto. *Iglesia Presbiteriana: A History of Presbyterians and Mexican Americans in the Southwest*. San Antonio: Trinity University Press, 1974.

Brogi, Andrea, Enrico Capezzuoli, Mehmet Cihat Alcicek, and Anna Gandin. "Evolution of a Fault-Controlled Fissure-Ridge Type Travertine Deposit in the Western Anatolia Extensional Province: The Çukurbağ Fissure-Ridge (Pamukkale, Turkey)." *Journal of the Geological Society* 171 (2014): 425–41.

Brown, David E. *The Grizzly in the Southwest: Documentary of an Extinction*. Norman: University of Oklahoma Press, 1996.

Bruni, Frank, and Elinor Burkett. *A Gospel of Shame: Children, Sexual Abuse, and the Catholic Church*. New York: Harper Perennial, 2022.

Bullock, Alice. *The Squaw Tree: Ghosts and Mysteries of New Mexico*. Santa Fe: Lightning Tree Press, 1978.

Cather, Willa. *Death Comes for the Archbishop*. Lincoln: University of Nebraska Press, 1999.

Chavez, Fray Angelico. "The Penitentes of New Mexico." *New Mexico Historical Review* 29, no. 2 (1954): 97–123.

Coop, Jonathan D., and Thomas J. Givnish. "Spatial and Temporal Patterns of Recent Forest Encroachment in Montane Grasslands of the Valles Caldera, New Mexico, USA." *Journal of Biogeography* 34, no. 5 (2007): 914–27.

Corral, Patricia G. "Bark Substance Utilization in the Warm Springs Area: A Northern New Mexico Example." Master's thesis, New Mexico Highlands University, 1996.

Crown, Patricia L., and W. Jeffrey Hurst. "Evidence of Cacao Use in the Prehispanic American Southwest." *Proceedings of the National Academy of Sciences* 106, no. 7 (2009): 2110–13.

References

Curry, Andrew. "Horse Nations: After the Spanish Conquest, Horses Transformed Native American Tribes Much Earlier Than Historians Thought." *Science* 379 (2023): 1288–93.

Davis, Douglas F., and L. V. Slonaker. "First Forest Service Wireless: Primary Sources in Forest History." *Journal of Forest History* 18, no. 1–2 (1974): 22–27.

deBuys, William. "Fractions of Justice: A Legal and Social History of the Las Trampas Land Grant, New Mexico." *New Mexico Historical Review* 56, no. 1 (1981): 71–97.

———, ed. *Seeing Things Whole: The Essential John Wesley Powell.* Washington, DC: Island Press, 2004.

Dewar, Jacqueline J., Donald A. Falk, Thomas W. Swetnam, Christopher H. Baisan, Craig D. Allen, Robert R. Parmenter, Ellis Q. Margolis, and Erana J. Taylor. "Valleys of Fire: Historical Fire Regimes of Forest-Grassland Ecotones across the Montane Landscape of the Valles Caldera National Preserve, New Mexico, USA." *Landscape Ecology* 36 (2021): 331–52.

Ebright, Malcolm. *Land Grants and Lawsuits in Northern New Mexico.* Guadalupita, NM: Center for Land Grant Studies, 2008.

Elliott, Michael L. *Large Pueblo Sites Near Jemez Springs, New Mexico.* Cultural Resources Report 3. Albuquerque: Southwestern Region, US Forest Service, 1982.

———. San José de los Jémez Mission and Gíusewa Pueblo Site, National Historic Landmark nomination form, National Park Service, 2011.

Elston, W. E. *Summary of the Mineral Resources of Bernalillo, Sandoval, and Santa Fe Counties, New Mexico.* Bulletin 81. Socorro: New Mexico Bureau of Mines and Mineral Resources, 1967.

Farella, Joshua, and Thomas W. Swetnam. "Terminus Ante Quem Dating of the Depopulation of Jemez Ancestral Villages." *Archaeology Southwest Magazine* 30, no. 4 (Fall 2016): 12–14.

Fink, J. H., and C. R. Manley. "Explosive Volcanic Activity Generated from within Advancing Silicic Lava Flows." In *Volcanic Hazards*, edited by J. H. Latter, 169–79. Berlin: Springer-Verlag, 1989.

Fliedner, Dietrich. "Pre-Spanish Pueblos in New Mexico." *Annals of the Association of the American Geographer* 65, no. 3 (1975): 363–77.

Flint, Richard, and Shirley Cushing Flint. 2009. "Bartolomé de Ojeda." State Records Center and Archives. http://www.newmexicohistory.org/filedetails.php?fileID=48. Accessed November 1, 2023.

Ford, Richard I. "The Cultural Ecology of Jemez Cave." In *From Mountain Top to Valley Bottom: Understanding Past Land Use in the Northern Rio Grande, New Mexico*, edited by Bradley J. Vierra, 69–79. Salt Lake City: University of Utah Press, 2013.

Forthofer, Jason M. "Fire Tornadoes." *Scientific American* 321, no. 6 (2019): 60–67.

Fowler, Catherine S. "Some Lexical Clues to Uto-Aztecan Prehistory." *International Journal of American Linguistics* 49, no. 3 (1983): 224–57.

Fowler, Don D. *The Western Photographs of John K. Hillers: Myself in the Water*. Washington, DC: Smithsonian Institution Press, 1989.

Ghattas, Monika W. *Los Árabes of New Mexico: Compadres from a Distant Land*. Santa Fe: Sunstone Press, 2012.

Glover, Vernon J. *Jemez Mountains Railroads: Santa Fe National Forest, New Mexico*. Albuquerque: Southwestern Region, US Forest Service, 1990.

Goff, Fraser, Jamie N. Gardner, Jeffrey B. Hulen, Dennis L. Nielson, Robert Charles, Giday WoldeGabriel, Francois-D. Vuataz, John A. Musgrave, Lisa Shevenell, and B. M. Kennedy. "The Valles Caldera Hydrothermal System, Past and Present, New Mexico, USA." *Scientific Drilling* 3 (1992): 181–204.

Goff, Fraser, Jamie N. Gardner, Steven L. Reneau, Shari A. Kelley, Kirt A. Kempter, and John R. Lawrence. *Geological Map of the Valles Caldera, Jemez Mountains, New Mexico*. Socorro: New Mexico Bureau of Geology and Mineral Resources, 2011.

Goff, Fraser, and Cathy J. Goff. *Overview of the Valles Caldera (Baca) Geothermal System*. Socorro: New Mexico Bureau of Geology and Mineral Resources, 2017.

Goff, Fraser, and Lisa Shevenell. "Travertine Deposits of Soda Dam, New Mexico, and Their Implications for the Age and Evolution of the Valles Caldera Hydrothermal System." *Geological Society of America Bulletin* 99, no. 2 (1987): 292–302.

Goff, Fraser, Lisa Shevenell, Jamie N. Gardner., Francois-D Vautaz, and Charles O. Grigsby. "The Hydrothermal Outflow Plume of Valles Caldera, New Mexico, and a Comparison with Other Outflow Plumes." *Journal of Geophysical Research* 93 (1988): 6041–58.

Gorenfeld, Will. "The Battle of Cieneguilla." *Wild West Magazine*, February 2008.

Guiterman, Christopher H., Rachel M. Gregg, Laura A. E. Marshall, Jill J. Beckmann, Phillip J. van Mantgem, Donald A. Falk, Jon E. Keeley, Anthony C. Caprio, Jonathan D. Coop, Paula J. Fornwalt, et al. "Vegetation Type Conversion in the US Southwest: Frontline Observations and Management Responses." *Fire Ecology* 18, no. 1 (2002): 1–16.

References

Guiterman, Christopher H., Ellis Q. Margolis, Craig D. Allen, Donald A. Falk, and Thomas W. Swetnam. "Long-term Persistence and Fire Resilience of Oak Shrubfields in Dry Conifer Forests of Northern New Mexico." *Ecosystems* 21 (2018): 943–59.

Guiterman, Christopher H., Ellis Q. Margolis, and Thomas W. Swetnam. "Dendroecological Methods for Reconstructing High-Severity Fire in Pine-Oak Forests." *Tree-Ring Research* 71, no. 2 (2015): 67–77.

Gutiérrez, Ramón A. "Aztlán, Montezuma, and New Mexico." In *Aztlán: Essays on the Chicano Homeland*, edited by Rudolfo, Anaya, Francisco A. Lomelí, and Enrique R. Lamadrid. Albuquerque: University of New Mexico Press, 2017.

Hagmann, R. Keala, Paul F. Hessburg, Susan J. Prichard, N. A. Povak, Peter M. Brown, Peter Z. Fulé, Robert E. Keane, Eric E. Knapp, J. M. Lydersen, K. L. Metlen, et al. "Evidence for Widespread Changes in the Structure, Composition, and Fire Regimes of Western North American Forests." *Ecological Applications* 31, no. 8 (2021): e02431.

Heintzman, Peter D., Grant D. Zazula, Ross D. E. MacPhee, Eric Scott, James A. Cahill, Brianna K. McHorse, Joshua D. Kapp, Mathias Stiller, Matthew J. Wooller, Ludovic Orlando, et al. "A New Genus of Horse from Pleistocene North America." *eLife* 6 (2017): e29944.

Hewett, Edgar L. "Antiquities of the Jemez Plateau, New Mexico." *Bulletin of the Smithsonian Institution, Bureau of American Ethnology* 32 (1906).

Holmes, William H. *Random Records of a Lifetime Devoted to Science and Art*, Vol. 5. Washington, DC: Smithsonian Institution, 1879–1894.

———. "Notes on the Antiquities of Jemez Valley, New Mexico." *American Anthropologist* 7, no. 2 (1905): 198–212.

Horgan, Paul S. *Lamy of Santa Fe*. Middletown, CT: Wesleyan University Press, 1975.

Huning, Franz. *Trader on the Santa Fe Trail: The Memoirs of Franz Huning*. Albuquerque: University of Albuquerque, 1973.

Isaacs, Judith A., ed. *Jemez Mountains 100 Years Ago: Stories Told by Joseph E. Routledge*. Jemez Pueblo: Butterfly & Bear Press, 2020.

Johnson, David M. *Final Report on the Battle of Cieneguilla: A Jicarilla Apache Victory Over the US Dragoons, March 30, 1854*. Albuquerque: Southwestern Region, US Forest Service, 2009.

Johnson, John W. *Reminiscences of a Forest Ranger, 1914–1944*. Dayton: OH: J. C. Robinette, 1976.

Julyan, Robert. *The Place Names of New Mexico.* Albuquerque: University of New Mexico Press, 1996.

Katzew, Ilona. *Casta Painting: Images of Race in Eighteenth-Century Mexico.* New Haven, CT: Yale University Press, 2004.

Kaye, Margot W., and Thomas W. Swetnam. "An Assessment of Fire, Climate, and Apache History in the Sacramento Mountains, New Mexico." *Physical Geography* 20, no. 4 (1999): 305–30.

Keleher, William A. *Turmoil in New Mexico.* Santa Fe: Rydal Press, 1952.

Kessell, John L. *Kiva, Cross & Crown: The Pecos Indians and New Mexico, 1540–1840.* Oro Valley: AZ: Western National Parks Association, 1995.

———. "A Long Time Coming: The Seventeenth-Century Pueblo-Spanish War." *New Mexico Historical Review* 86, no. 2 (2011): 141–56.

Kitzberger, Thomas, Peter M. Brown, Emily K. Heyerdahl, Thomas W. Swetnam, and Thomas T. Veblen. "Contingent Pacific-Atlantic Ocean Influence on Multi-Century Wildfire Synchrony over Western North America." *Proceedings of the National Academy of Sciences* 104, no. 2 (2007): 543–48.

Kitzberger, Thomas., Donald A. Falk, Anthony L. Westerling, and Thomas W. Swetnam. "Direct and Indirect Climate Controls Predict Heterogeneous Early–Mid 21st Century Wildfire Burned Area across Western and Boreal North America. *PloS One* 12, no. 12 (2017): e0188486.

Kues, Barry S. "Guide to the Late Pennsylvanian Paleontology of the Upper Madera Formation, Jemez Springs Area, North-Central New Mexico." In *Jemez Mountains Region*, edited by F. Goff., B. S. Kues, M. A. Rogers, L. S. McFadden, and J. N. Gardner, 169–88. Socorro: New Mexico Geological Society, 1996.

Kulisheck, Jeremy. "The Archaeology of Pueblo Population Change on the Jemez Plateau, AD 1200 to 1700: The Effects of Spanish Contact and Conquest." PhD dissertation, Southern Methodist University, 2005.

Kulisheck, Jeremy, Anastasia Steffen, and Connie Constan. "Agricultural Intensification, Field Features, and Hypervisibility: A Case from the Jemez Mountains, New Mexico, USA." Poster paper presented at Seventy-Eighth Annual Meeting of the Society for American Archaeology, Honolulu, 2013.

LeCompte, Mary Lou. "The Hispanic Influence on the History of Rodeo, 1823–1922." *Journal of Sport History* 12, no. 1 (1985): 21–38.

Leopold, Aldo. *A Sand County Almanac and Sketches Here and There.* New York: Oxford University Press, 1968.

References

Liebmann, Matthew. *Revolt: An Archaeological History of Pueblo Resistance and Revitalization in 17th Century New Mexico.* Tucson: University of Arizona Press, 2012.

Liebmann, Matthew J., Joshua Farella, Christopher I. Roos, Adam Stack, Sarah Martini, and Thomas. W. Swetnam. "Native American Depopulation, Reforestation, and Fire Regimes in the Southwest United States, 1492–1900 CE." *Proceedings of the National Academy of Sciences* 113, no. 6 (2016): E696–704.

Lipe, William D., Shannon Tushingham, Eric Blinman, Laurie Webster, Charles T. LaRue, Aimee Oliver-Bozeman, and Jonathan Till. "Staying Warm in the Upland Southwest: A 'Supply Side' View of Turkey Father Blanket Production." *Journal of Archaeological Science Reports* 34 (2020): 102604.

Lucas, Spencer G., Susan K. Harris, Justin A. Spielmann, David S. Berman, Amy C. Henrici, Karl Krainer, Larry F. Rinehart, William A. Dimichele, Dan S. Chaney, and Han S. Kerp. "Lithostratigraphy, Paleontology, Biostratigraphy, and Age of the Upper Paleozoic Abo Formation near Jemez Springs, Northern New Mexico." *Annals of Carnegie Museum* 80, no. 4 (2012): 323–50.

Maingi, John K., and Thomas W. Swetnam. *Peeled Ponderosa Pine Trees from Northern New Mexico.* Tucson: Laboratory of Tree-Ring Research, University of Arizona, 1993.

Mann, Michael E. *Our Fragile Moment: How Lessons from Earth's Past Can Help Us Survive the Climate Crisis.* New York: Public Affairs, 2023.

Martin, Craig. *Valle Grande: A History of the Baca Location No. 1.* Los Alamos: All Seasons Publishing, 2003.

Martorano, Marilyn A., and J. Beardsley. "New Mexico's Forgotten Resource: Culturally Peeled Trees." In *Why Museums Collect: Papers in Honor of Joe Ben Wheat*, edited by Meliha Duran and David Kirkpatrick. Albuquerque: Archaeological Society of New Mexico, 1993.

Miller, Darlis A. *Matilda Coxe Stevenson: Pioneering Anthropologist.* Norman: University of Oklahoma Press, 2007.

Mohajjel, Mohammed. "Quaternary Travertine Ridges in the Lake Urmia Area: Active Extension in NW Iran." *Turkish Journal of Earth Sciences* 23 (2014): 602–14.

Momaday, N. Scott. *House Made of Dawn.* New York: Harper & Row, 1968.

Morgan, Gary S., and Spencer G. Lucas. "Pleistocene Vertebrate Faunas in New Mexico from Alluvial, Fluvial, and Lacustrine Deposits." In *New Mexico's Ice Ages*, edited by Spencer G. Lucas, Gary S. Morgan, and Kate E. Zeigler, 185–248. Albuquerque: New Mexico Museum of Natural History and Science, 2005.

Morgan, Gary S., and Larry F. Rinehart. "Late Pleistocene (Rancholabrean) Mammals from Fissure Deposits in the Jurassic Todlito Formation, White Mesa Mine, Sandoval County, North-Central New Mexico." *New Mexico Geology* 29, no. 2 (2007): 30–51.

Nasholds, Morgan W. M., and Matthew J. Zimmerer. "High-Precision 40Ar/39Ar Geochronology and Volumetric Investigation of Volcanism and Resurgence Following Eruption of the Tshirege Member, Bandelier Tuff, at the Valles Caldera." *Journal of Volcanology and Geothermal Research* 431 (2022): 107624.

Nelson, Kate. "Favorite Sun." *New Mexico Magazine*, January 8, 2019.

Niederman, Sharon. *A Quilt of Words: Women's Diaries, Letters & Original Accounts of Life in the Southwest 1860–1960*. Denver: Bower House, 1990.

Opler, Morris E. *Myths and Tales of the Jicarilla Apache Indians*. New York: Dover Publications, 2012.

Pettitt, Roland A. *Exploring the Jemez Country*. Los Alamos: Los Alamos Historical Society, 1990.

Pritchard, Susan J., Paul F. Hessburg, R. Keala Hagmann, Nicholas A. Povak, Solomon Z. Dobrowski, Matthew D. Hurteau, Van R. Kane, Robert E. Keane, Leda N. Kobziar, Crystal A. Kolden, et al. "Adapting Western North American Forests to Climate Change and Wildfires: 10 Common Questions." *Ecological Applications* 31, no. 8 (2021): e02433.

Read, Benjamin M. "The Last Word on Montezuma." *New Mexico Historical Review* 1, no. 3 (1926): 350–58.

Ritch, William G. *Aztlán: The History, Resources and Attractions of New Mexico*. Boston: D. Lothrop & Company, 1885.

Roberts, David. *The Pueblo Revolt: The Secret Rebellion that Drove the Spaniards Out of the Southwest*. New York: Simon & Schuster, 2005.

Robinson, William J., John W. Hannah, and Bruce G. Harrill. *Tree-ring Dates from New Mexico, I, O, U: Central Rio Grande Area*. Tucson: Laboratory of Tree-Ring Research, University of Arizona, 1972.

Rogers, Margaret A., Barry S. Kues, Fraser Goff, Frank J. Pazzaglia, Less A. Woodward, Spencer G. Lucas, and Jamie N. Gardner. "First-Day Road Log, from Bernalillo to San Isidro, Southern Jemez Mountains, Guadalupe Box, Jemez Springs, Valles Caldera, and Los Alamos." In *Jemez Mountains Region*, edited by F. Goff, B. S. Kues, M. A. Rogers, L. S. McFadden, and J. N. Gardner, 28. Socorro: New Mexico Geological Society, 1996.

References

Romer, Alfred S. "The Vertebrate Fauna of the New Mexico Permian." *New Mexico Geological Society Guidebook* 11 (1960): 48–54.

Roos, Christopher I. "The Long-Term Context for Human–Fire Relationships, on the Jemez Plateau." *Archaeology Southwest Magazine* 30, no. 4 (2016): 21–22.

Roos, Christopher I., and Christopher H. Guiterman. "Dating the Origins of Persistent Oak Shrubfields in Northern New Mexico Using Soil Charcoal and Dendrochronology." *Holocene* 31, no. 7 (2021): 1212–20.

Roos, Christopher I., Christopher H. Guiterman, Ellis Q. Margolis, Thomas W. Swetnam, Nicholas C. Laluk, Kerry F. Thompson, Chris Toya, Calvin A. Farris, Peter Z. Fulé, Jose M. Iniguez, et al. "Indigenous Fire Management and Cross-Scale Fire-Climate Relationships in the Southwest United States from 1500 to 1900 CE." *Science Advances* 8 (2022): eabq3221.

Roos, Christopher I., Tammy M. Rittenour, Thomas W. Swetnam, Rachel A. Loehman, Kacy L. Hollenback, Matthew J. Liebmann, and Dana Drake Rosenstein. "Fire Suppression Impacts on Fuels and Fire Intensity in the Western US: Insights from Archaeological Luminescence Dating in Northern New Mexico." *Fire* 3, no. 3 (2020): doi:10.3390/fire3030032.

Roos, Christopher I., and Thomas W. Swetnam, eds. 2016. "Fire Adds Richness to the Land: The Jemez FHiRE Project." *Archaeology Southwest Magazine* 30, no. 4 (2016).

Roos, Christopher I., Thomas W. Swetnam, T. J. Ferguson, Matthew J. Liebmann, Rachel A. Loehman, John R. Welch, Ellis Q. Margolis, Chris H. Guiterman, William C. Hockaday, Michael J. Aiuvalasit, et al. "Native American Fire Management at an Ancient Wildland–Urban Interface in the Southwest United States." *Proceedings of the National Academy of Sciences* 118, no. 4 (2021): e2018733118.

Rowe, T. B., T. W. Stafford Jr., D. C. Fisher, J. J. Enghild, J. M. Quigg, R. A. Ketcham, J. C. Sagebiel, R. Hanna, and M. W. Colbert. "Human Occupation of the North American Colorado Plateau ~37,000 Years Ago." *Frontiers in Ecology and Evolution* 10 (2022): doi.org/10.3389/fevo.2022.903795.

Sanchez, Frederick R. "The Presbyterians in the Valley of the Holy Eucharist: The Presbyterian Missionary Movement in the Late 19th and Early 20th Centuries in Jemez, New Mexico," no date. Menaul Historical Library, Albuquerque.

Sando, Joe S. *Nee Hemish: A History of Jemez Pueblo.* Santa Fe: Clear Light Publishing, 2008.

Scanland, J. M. "Theft of the Sacred Fire." *Topeka Daily Herald*, November 1, 1902.

Schlaepfer, Martin A., Dov F. Sax, and Julian D. Olden. "The Potential Conservation Value of Non-Native Species." *Conservation Biology* 25, no. 3 (2011): 428–37.

Scurlock, Dan. "Euro-American History of the Study Area." In *High Altitude Adaptations along Redondo Creek*, edited by Craig Baker and Joseph C. Winter, 131–60. Albuquerque: Office of Contract Archeology, University of New Mexico, 1981.

———. *From the Rio to the Sierra: An Environmental History of the Middle Rio Grande Basin*. Fort Collins, CO: US Forest Service, 1998.

Sheppard, Paul R., Arnie Friedt, Mike Neathamer, and Furrukh Bashir. "Dendrochronology of Culturally Modified Trees, Philmont Scout Ranch, NM." Unpublished poster paper. Tucson: Laboratory of Tree-Ring Research, University of Arizona, 2016.

Simpson, James H. *Navaho Expedition: Journal of a Military Reconnaissance from Santa Fe, New Mexico, to the Navaho Country Made in 1849*. Edited and annotated by Frank McNitt. Norman: University of Oklahoma Press, 1964.

Stegner, Wallace. *Beyond the Hundredth Meridian: John Wesley Powell and the Second Opening of the West*. Lincoln: University of Nebraska Press, 1982.

Stromberg, Julie. "Dynamics of Fremont Cottonwood (*Populus fremontii*) and Saltcedar (*Tamarix chinensis*) Populations along the San Pedro River, Arizona." *Journal of Arid Environments* 40, no. 2 (1998): 135–55.

Swetnam, Thomas W. "Peeled Ponderosa Pine Trees: A Record of Inner Bark Utilization by Native Americans." *Journal of Ethnobiology* 4, no. 2 (1984): 177–90.

———. "Tree-ring Dates from the Hot Sulphur Bath House, Jemez Springs, New Mexico." Jemez Valley History, 2015. https://jemezvalleyhistory.org/wp-content/uploads/2024/08/HotSulphurBathHouseReport_111715_Secure.pdf.

———. "Fire History of a Ponderosa Pine Stand, Area 3, San Diego Canyon, Jemez Ranger District, Santa Fe National Forest." Jemez Valley History, 2017. https://jemezvalleyhistory.org/wp-content/uploads/2024/08/Area3_FireHistory_April2017_Secure.pdf.

———. "Late Nineteenth and Early Twentieth Century Logging in the Jemez Valley: Log Chutes in Area 3." Jemez Valley History, 2017. https://jemezvalleyhistory.org/wp-content/uploads/2024/08/LogChutesArea3_Report_Secure.pdf.

———. "Peeled Ponderosa Pine Trees at the Warm Springs, Battle of Cieneguilla Site, New Mexico." Jemez Valley History, 2017. https://jemezvalleyhistory.org/wp-content/uploads/2024/08/WarmSpringsPeeledTrees_ReportAndApp_061817_Secure.pdf.

———. "Tree-Ring Dates of the Stone Hotel." Jemez Valley History, 2019. https://jemezvalleyhistory.org/wp-content/uploads/2024/08/StoneHotel_Bodhi_TreeRingDates_Revised_091622_Secure.pdf.

Swetnam, Thomas W., Craig D. Allen, and Julio L. Betancourt. "Applied Historical Ecology: Using the Past to Manage for the Future." *Ecological Applications* 9, no. 4 (1999): 1189–206.

Swetnam, Thomas W., and Julio L. Betancourt. "Fire-Southern Oscillation Relations in the Southwestern United States." *Science* 249 (1990):1017–20.

———. "Mesoscale Disturbance and Ecological Response to Decadal Climatic Variability in the American Southwest." *Journal of Climate* 11 (1998): 3128–47.

Swetnam Thomas W., Joshua Farella, Christopher I. Roos, Matthew J. Liebmann, Donald A. Falk, and Craig D. Allen. "Multi-Scale Perspectives of Fire, Climate, and Humans in Western North America and the Jemez Mountains, USA." *Philosophical Transactions of the Royal Society* B 371 (2016): 20150168.

Swetnam, Tyson L., Ann M. Lynch, Donald A. Falk, Steve R. Yool, and D. Philip Guertin. "Discriminating Disturbance from Natural Variation with LiDAR in Semi-arid Forests in the Southwestern USA." *Ecosphere* 6, no. 6 (2015): 97.

Tafoya, Jean. "Uranium-Series Geochronology and Stable Isotope Analysis of Travertine from Soda Dam, Mew Mexico: A Quaternary Record of Episodic Spring Discharge and River Incision in the Jemez Mountains Hydrothermal System." Master's thesis, University of New Mexico, 2012.

Taylor, William T. T., Pablo Librado, Mila Hunska Tašunke Icu, Carlton Shield Chief Gover, Jimmy Arterberry, Anpetu Luta Win, Akil Nujipi, Tanka Omniya, Mario Gonzalez, Bill Means, et al. "Early Dispersal of Domestic Horses into the Great Plains and Northern Rockies." *Science* 379 (2023):1316–23.

Tosa, Paul. "Battle of Astialakwa." *Walatowan*, October–November 2021.

Tosa, Paul, Matthew J. Liebmann, T. J. Ferguson, and John R. Welch. "Movement Encased in Tradition and Stone: Hemish Migration, Land Use, and Identity." In *The Continuous Path: Pueblo Movement and the Archaeology of Becoming*, edited by Samuel Duwe and Robert W. Preucel, 60–77. Tucson: University of Arizona Press, 2019.

Towner, Ronald H., and Stacey K. Galassini. "Cambium-Peeled Trees in the Zuni Mountains, New Mexico." *Kiva* 78, no. 2 (2012): 207–27.

Townshend, Richard B. *A Tenderfoot in Colorado*. Norman: University of Oklahoma Press, 1968.

———. *Last Memories of a Tenderfoot*. 1926. Reprint, Whitefish, MT: Kessinger, 2010.

———. *A Tenderfoot in New Mexico*. 1923. Reprint, Santa Fe: Sunstone Press, 2013.

Tucker, Edwin A. *The Early Days: A Sourcebook of Southwestern Region History*. Book 2, *Cultural Resources Management*. Albuquerque: Southwestern Region, US Forest Service, 1991.

Tucker, Edwin A., and George Fitzpatrick. *Men Who Matched the Mountains: The Forest Service in the Southwest*. Albuquerque: Southwestern Region, US Forest Service, 1974.

Twitchell, Ralph Emerson. *Old Santa Fe: A Magazine of History, Archaeology, Genealogy, and Biography*. Santa Fe: Old Santa Fe Press, 1912–1914.

Van Hart, Dirk. *Camps and Campsites of the Civilian Conservation Corps in New Mexico, 1933–1942*. Santa Fe: Sunstone Press, 2020.

Van Valkenburgh, Richard. "Astaelakwa, House of the Vanished." *Desert Magazine*, April 1942.

Weber, David J. *What Caused the Pueblo Revolt?* Boston: Bedford/St. Martins, 1999.

Webster, J. Shannon. "José Martínez and the New Mexico Reformation." Undated, unpublished manuscript shared by the author.

Westerling, Anthony L., Hugo G. Hidalgo, Daniel R. Cayan, and Thomas W. Swetnam. "Warming and Earlier Spring Increase Western US Forest Wildfire Activity." *Science* 313, no 5789 (2006): 940–43.

Wheeler, George M. *Explorations and Surveys West of the One Hundredth Meridian*. Vol. 3, *Geology*. Washington, DC: US Government Printing Office, 1875.

———. "US Geographical Surveys West of the One Hundredth Meridian, Parts of Southern Colorado and Northern New Mexico." Atlas sheet 69. American Geographical Society Library, University of Wisconsin, Milwaukee.

White, Leslie A. "The Pueblo of Sia, New Mexico." *Smithsonian Institution, Bureau of American Ethnology Bulletin* 184 (1962).

Wiegner, Kathleen, and Robert Borden. *Jemez Springs*. Dover, NH: Arcadia Publishing, 2009.

Wilgus, Justin, Brandon Schmandt, Ross Maguire, Chengxin Jiang, and Julien Chaput. "Shear Velocity Evidence of Upper Crustal Magma Storage beneath Valles Caldera." *Geophysical Research Letters* 50 (2023): e2022GL101520.

Williams, A. Park, Craig D. Allen, Alison K. Macalady, Daniel Griffin, Connie A. Woodhouse, David M. Meko, Thomas W. Swetnam, Sara A. Rauscher, Richard Seager, Henri D. Grissino-Mayer, et al. "Temperature as a Potent Driver of Regional Forest Drought Stress and Tree Mortality." *Nature Climate Change* 3, no. 3 (2013): 292–97.

References

Williams, A. Park, Benjamin I. Cook, and Jason E. Smerdon. "Rapid Intensification of the Emerging Southwestern North American Megadrought in 2020–2021." *Nature Climate Change* 12 (2022): 232–34.

Williams, A. Park, Edward R. Cook, Jason E. Smerdon, Benjamin I. Cook, John T. Abatzoglou, Kasey Bolles, Seung H. Baek, Andrew M. Badger, and Ben Livneh. 2020. "Large Contribution from Anthropogenic Warming to an Emerging North American Megadrought." *Science* 368 (2020): 314–18.

Williams, A. Park, Richard Seager, Max Berkelhammer, Alison K. Macalady, Micahael A. Crimmins, Thomas W. Swetnam, Anna T. Trugman, Nikolaus Buenning, Natalia Hryniw, Nate G. McDowell, et al. "Causes and Implications of Extreme Atmospheric Moisture Demand during the Record-Breaking 2011 Wildfire Season in the Southwestern United States." *Journal of Applied Meteorology and Climatology* 53, no. 12 (2014): 2671–84.

Zimmerer, Mathew J., John Lafferty, and Matthew A. Coble. 2016. "The Eruptive and Magmatic History of the Youngest Pulse of Volcanism at the Valles Caldera: Implications for Successfully Dating Late Quaternary Eruptions." *Journal of Volcanology and Geothermal Research* 310 (2016): 50–57.

INDEX

Abousleman: family, history of, 59; family moves to Jemez Hot Springs, 59; fire in the home of, 290n7; Fred, 64; Leonard Lewis visit to sawmill, 90; mercantile converted to Los Ojos Bar, 1947, 65; McCauley Warm Springs, 269; Moses, 59, 64, 176, 290n7, 290n9; Moses ends partnership with Salmon, 61; Moses finds underwater passage in Soda Dam, 177; Moses owner of 40,000 sheep in 1910, 61; Moses purchase of Judt Hot Mineral Bath house, 61; Moses shot by sheep rustlers, 62; Moses steadfastly refuses arm amputation, 64; Moses wounded by rustlers, 62; photograph of mercantile with Jemez Pueblo dancers, 43; store robberies, 64

Abrego, Manuel, 112

acequia waters, 6–7

Acoma, Aztecs visit to, 44

Adams: family, 87; Cañon de San Diego Grant document written by, 72, 291n6; John A. Jr., 72; John Amos Sr., 85, 87, 292nn3–4; John Amos Sr., married Helen Shields, 72. *See also* Shields family

Adams, Token, 94

agricultural fields, mapping of Hemish, by Dietrich Fliedner, *Pre-Spanish Pueblos in New Mexico*, 312n2

Alexander, Hubert G., 314n5, 314n7, 318; *The Excavation of Jemez Cave*, 175, 301n3; Jemez Cave photograph 1935, 177; statement about workmen camping in cave beneath Soda Dam, 300n3, 303n2

Allen, Craig D., op-ed in *Santa Fe New Mexican* on wildfires, 219, 309n1; quoted in *New York Times* on Las Conchas Fire, "As Fires Grow, a New Landscape Appears in the West," 307n9; study of montane grasslands, 213–14

Apache: fighting with pueblos, 19; place names and stories, 277; raiding by, and also Ute and Navajo, 127

archbishop: Gerald Fitzgerald letters to Edwin Byrne, 265; Jean-Baptiste Lamy, 127

Archuleta: Andres, 136; Calletana, married James Smith, 249; Emiterio (Miterio), 112–18, 131–33, 142; José Francisco, 61, 112, 116, 131–34, 136, 297n6, 298n2, 299n3; Manuel, 61; Maria Viviana Montoya, 112, 131, 134, 249; two sons killed by Navajos, 109

Astialakwa: Battle of, 20, 285n11, 287n12; infiltrators during, 286n8; map of location, 21

Atchison, Topeka & Santa Fe (AT&SF) railway: arrival in New Mexico, 77; building hotels, 243; rail line planned to Jemez Hot Springs, 243

Aztecs: Montezuma chief of, 39; Zia and Jemez Pueblo contact with, pre-Spanish, 43

Aztlán: 1885 book by William G. Ritch, 41; more of a concept than a place, 42

Baca Location No. 1, 84; logging by NML&T, 83; Mariano Otero ownership of, 72; mortgage held by Perea, 129; number of livestock grazed there in 1918, 310n4; wildfires fought there by Forest Service, 88

Baca Ranch, 95

BAE. *See* Bureau of American Ethnology

Index

Banco Bonito, 165, 301n6; forest in 1904, 201–2; lidar map of, 178; Tenth Circuit Court ruling regarding aboriginal title, 71, 272; lava eruption vent, flows, and age of, 179, 301n6

Bandelier, Adolf F.: description of artifacts at San José de los Jémez mission in 1891, 285n11; scholar of Spanish and Mexican documents, 39; statement about defensive aspect of San José de los Jémez, 16; theory of nineteenth-century Montezuma myths, 39–40

Bandelier Tuff, upper, 194

Barbour, Matthew, *Early Farmers of Jemez Cave*, 314n5; regarding 1623 uprising, 284n5

Barley Cañon, 142

bathhouse: 1880 newspaper article about the construction of, 244; 1884 photograph of, 134; Archuleta log cabin, 297n6; owned by Judt before Abousleman, 61; Perea moved to Jemez Hot Springs to manage in 1881, 77; stone-walled built by Mariano Otero and Francisco Perea, 113, 129; at Sulphur Springs, burned down in the 1970s, 52

battles: of 1689, 1694, and 1696, 17–24; of 1689 at Zia, 18; of 1694, 22; of 1694 and "burros could be heard braying like coyotes," 287n12; accuracy of 1689 account questioned, 285n5; of Astialakwa one day before feast of Santiago, 286n9; refugees from 1694 and 1696, 23

Battle of Cieneguilla, 1854 battle between US Army and Jicarilla Apaches, 100, 294n8

Battleship Campground, improvements and octagonal stone pavilion built by CCC, 253

Battleship Rock, 179, 181; 1933 wildfire at, 312n6; Boy Scout camp at, 312n5; James Smith homestead and mill at, 312n4; co-magmatic with El Cajete, 301n6; drawing of, 241; Dietrich Fliedner mapping of archaeological sites near, 312n2; formation of, 165, 181, 249; La Junta Fly Camp at, 252–53; UNM archaeology field camp at, 251

bear: grizzly, sleeping at Soda Dam, 103; grizzly, encounter by William Henry Holmes, 101–3; grizzly and Routledge milk cow, 104; last grizzly in New Mexico, 105; woman runner attacked by black bear in Valles Caldera, 296n1

bell tower, octagonal, of San José de los Jémez mission: kids climbing on, 16; photographs of, 13

Berger, Otto, 69; case car agent, of first automobile driven to Jemez Hot Springs in 1912, 68

Betancourt, Julio, 214

Bland, New Mexico: ranger station at, 87; mining company store in, 104; stage line to Jemez through, 244

blasting of Soda Dam in 1961, 167–71; NM Department of Transportation role in, 167

Block, John B., hotelier in Jemez Springs, 74

Bloom, Lansing, 319; description of abandoned dugout log on Borrego Mesa, 6; review of early Spanish visits to Jemez, 3; statement about San José de los Jémez church, 284n1

Bohdi Manda Zen Buddhist center (originally Stone Hotel), 134, 129, 244

Borrego Mesa, 6

Boy Scouts: buildings and 1926 camp at Battleship Rock, 251–52, 312n5; planted Russian olive trees along Jemez River, 229

bridge, photographs of, below Soda Dam before 1941 washout and modern, 186

Brown, Ron, 207

Bureau of American Ethnology: John K. Hillers, photographer of, 33; John Wesley Powell, first director of, 37
Burnett Mining Company, 28
burning, prescribed: need for, 204; positive effects of, 149
Bursum, Holm Senator, passed *Pueblo Lands Condemnation Act*, 78
Byrne, Edwin V., Archbishop of Santa Fe, letter from Father Fitzgerald to, 264

Cabeza de Baca, Luis Maria, original recipient of 1860 Baca Location grant, 71
Cabot, Bruce, visit to Rowland's ranch in 1938, 259
Calaveras Canyon, peeled trees at, 96–97
Caldwell: Lew, 90; Mary (née Fenton), 131
cambium-peeled trees, 95–100, 294nn7–9
Campbell, Jack, Governor, 16
Camp Kearny, 257; CCC camp at Battleship Rock, 313n8
Camp Porter, 257; Boy Scout camp at Battleship Rock, 251; logging camp on Rio Guadalupe, 74
Camp Shaver: ARC sessions, 257; area, 312n2; buildings and barracks cabins at, 254; camp counselors, 254–56, 276; National Forest and land grant boundary, 250; mess hall, 252–54; photograph of staff members in 1965, 255
canoa (or conovas): dugout log for acequia, 6–7
Cañoncito, 112; also called Cañon, 115; at junction of Rios Guadalupe and Jemez, 109
Canovas Canyon, 6
carbon-14 dating: of charcoal from wildfires and plant fossils, 182, 199; of child burial from Jemez Cave, 261; of corncobs from Jemez Cave, 155; of horse bones, 194

Carleton, James General, 99
Carson, Kit Colonel, defeat of Navajos in 1864, 110, 297n6
Carson National Forest, 85, 92
Casados, Sigfredo, 64
Cat Mesa: forest thinning on, 149; Oñate travels across in 1598, 5; Padre Alonzo Trail on, 52; photograph of 2022 Cerro Grande Fire on, 216
Cather, Willa, *Death Comes for the Archbishop*, 127
cave: beneath Soda Dam used for camping in 1934–1935, 174, 301n3; underwater passage in Soda Dam found by Moses Abousleman, 176; hidden beneath northside of Soda Dam, 176–77; in volcanic tuff used by José Francisco Archuleta (or Sandovals), 297n6, 299n3
CCC. *See* Civilian Conservation Corps
Cebolla Creek. *See* Rio Cebolla
Cebollita Fire, 205–10; Forest Supervisor describes, 207; in *Life Magazine*, July 1971, 207; "monstrous serpent writhing," 207; photographs of, 206; re-forestation of burn scar, 209–10
census, precincts 10 and 9 in 1920 and 1930, 80, 293n6
Cerro Colorado, refugee mesa pueblo of Zia people after 1689 battle, 19
Cerro Grande, 237; 2000 wildfire and investigation report, 211, 309n8; grassland, photographs of, 212–13; paleo-history of, 213; restoration of grassland, 214
Cerro Pelado Fire, 149; burn severity map, 218, 219; post-wildfire effects on erosion, 220; probable cause of, 212, 219; view from Area 3, April 22, 2022, 216
Cerro Pelado Peak, name origin and lookout tower, 211
Chaco Canyon: pueblo trade with meso-Americans, 42; Wetherills at, 88

Index

Chama River: Jemez refugees at, 22; part of Jemez National Forest, north of, 85; Permian fossils in red rocks at, 29; Pleistocene fossils found near, 193

Charles, Perl, 86, 88–89; first district ranger on Jemez River District, 87; story of livestock impoundment confrontation near Coyote, New Mexico, 88

Chavez, Amado, partner of Alonzo McMillen in New Mexico land grant takings, 73

Church Canyon: acequia flume across, 7; first ranger station at, 87; in Hiller's 1879 photograph, 37; Juan de Oñate travels down in 1598, 5; Leonard Lewis rides up, 90; location of Gíusewa, 9; Padre Alonzo Trail in, 52

Cieneguilla, Battle of, peeled trees near, 100

Civilian Conservation Corps, 253, 257, 313n8; La Junta Fly Camp, 252; reconstruction of bell tower staircase in mission San José de los Jémez, 13, 284n9

Civil War, 37, 52, 127, 129

climate oscillations, 237–38; ocean-atmosphere, ENSO, 238

climate change: in the Jemez, 231; skepticism of, 237; and wildfires, 214, 221

Coloradas, Mangus, Mimbreño Apache leader, 99

Comanche raiding, 297n4

copper, at Spanish Queen Mine, 27–28, 191

Correr El Gallo, 120–22, 298n4

Cortés, Hernán, 39

cottonwood trees: 197, 223, 229; bosque expansion, 223; floods and regeneration dynamics of, 227, 309n3; and junipers, 223; reasons for absence in late 1800s, 227

coyote: mixed heritage class name, also known as mestizo, 17, 285n4; petroglyph on boulder near Guadalupe Mesa, 287n12

Coyote Pass, saddle between Guadalupe and Virgin Mesas, 23

craters, lava explosion pits on Banco Bonito, 181. *See also* explosion craters

Cruiser Rock, 256

culturally modified, peeled, or carved-upon trees, 95–98, 294n7. *See also* inner bark

Curry, Jim, 86, 88

Dancing Coyotes, enigmatic petroglyph on boulder near Guadalupe Mesa, 287n12

dates: of Banco Bonito lava flows, 179; photograph of fires on tree-ring cross-section, *198*; of Soda Dam blasting, 302n1; of Soda Dam formation, 161; tree-ring, 1810 of peeled trees in Calaveras Canyon, 96; tree-ring, of Hemish ancestral village depopulation, 203; tree-ring, of house built within convento of San José de los Jémez, 284n8; tree-ring, of peeled trees in New Mexico, 294n1

de Arvide, Martin, directed final reconstruction of San José de los Jémez, 15; forces Hemish to resettle Gíusewa, 284n5; third priest at San José de los Jémez, 11

de Benavides, Alonso, 11, 283n1

deBuys, William, photograph in History Grove, *98*

de Cruzate, Domingo Jironza Petrís, 18

de Lugo, Alonzo, 9

de Ojeda: Bartolomé, 17–20, 285n3–4, 286n6–7; his "conversion" according to Flint, 286n6; as a mixed-blood

("coyote"), 21, 287n12; wounded in 1689 battle, 18–19
de Oñate, Juan, 5, 109
de Vargas, Diego, 19–20, 22, 24
Dimetrodon occidentalis, fossils of, 29, 31, 191
district ranger, 88–89; first on Jemez River Ranger District, 272
drought: of the 1580s, 235; of the 2000s (most extreme in 1,200 years), 235, 309n1; of the past millennium, 234; Southwest is ground zero of, 231; temperature as a potent driver of, 233–35, 310n2–3, 311n9

eagle: and cactus bush at Mexico City legend, 42; image on Redondo Peak (Wâavêmâ), 211, 308n3; Montezuma flying on the back of, 41–42, 42
East Fork: dammed multiple times by lava and pyroclastic flows, 179, 249; blocked by South Mountain lava flow, 165; possible location of Powell-proposed dam at Las Conchas, 57
El Cajete: and Battleship Rock pyroclastic flows, 165, 181, 301n6; lava vent on Banco Bonito, 249
Elias, Vicente: accompanied R. B. Townshend to Jemez in 1874, 112; his murderers expiating their sins, 117; murdered, 112
Elliott, Michael: about Jemez ancestral village of Pejunkwa, 283n6; National Historic Landmark form, 296n1; quote from Bandelier's visit to San José de los Jémez mission, 285n11
El Niño-Southern Oscillation (ENSO), 237–38
Emerson, Ralph Waldo, "He builded better than he knew;— / The conscious stone to beauty grew," 47, 52

Equus francisci (re-named *Harringtonhippus francisci*) fossil, stilt-legged horse from White Mesa Mine, 192, 194
ethnology: Bandelier's studies in, 58; subfield of anthropology, 37
excavation, of Jemez Cave, 164, 177; Alexander, *The Excavation of Jemez Cave*, 300n3
expedition: 1879 BAE, 33; 1887 Powell, Stevenson, Holmes, et al. in the Jemez, 37
explosion craters: of Banco Bonito and map, 178–80; phreatic type of, 305n7; around Redondo campground, map of, 181; volatile gases resulting in, 181, 279
extinction of grizzly bears in New Mexico, 105

FDSI. *See* Forest Drought Stress Index
feathers, turkey, used to make blankets, 26. *See also* turkey
Feit, John Bernard, 267
Fenton: E. M. (Mac) Jr., 131, 144, 299n3; Elijah Maclean Sr., 138, 140, 143; Elijah Maclean Sr., about his marriage, 142; Elijah Maclean Sr., about prairie dogs, 141; Jean, 139; Jessi, 131, 137, 139; Joshua Maclean, 131; visit to the family on the Rio Cebolla in 1903, 131
Fenton Lake, 131, 137
Fettes, Anthony, 173
Field, Matthew C., *Sacred Fire* story from Pecos, told in 1839, 289n5
field houses, Hemish ancestral, 249, 257, 312n2
fires: on the Baca Location and San Diego Grant, 88; in cottonwood bosque, 223; history of, 199; low-severity, 182; suppression of, 203; wind-driven, 207

335

Index

firewood, large quantities used, 203
fissure ridge travertine deposit, at Soda Dam, 155, 176
fissure, central, in Soda Dam, 155–56, 162–67, 169, 171, 173; in White Mesa gypsum mine, 193
Fitzgerald, Gerald Michael Cushing, 263–65, 267–68, 314n3, 315n5, 316n12, 317n13; co-founder of Servants of the Paraclete, 263; letters to Catholic leadership on pedophilic priests, 264–65
flagellant, orders, 116, 133
fleur-de-lis, wall decorations at San José de los Jémez mission church, 15
floods: 1874 event, 184; 1886 unprecedented freshet, 48; 1924 washing out part of SFNW rail line, 80; 1941 damage to Abousleman bath house, 61; 1941 event estimated flow level, 305n1; 1941 over-topping of Soda Dam, 157, 176; 1941 washing out of stuck log at Soda Dam, 154; 1958 washing out of west wing of Stone Hotel, 183, 189; catastrophic, in San Diego Canyon, 165; in Church Canyon, 33; cottonwood trees establishing after 1941 and 1958 floods, 227; great event at Soda Dam of centuries ago, 176; list of large events on Jemez River, 183–85, 189, 227; photograph of effects of 1941 event below Guadalupe box, 228
Fluke: George, 187; Harry, 187
flume, over Church Canyon arroyo, 7
forest conversion to shrublands, 209, 307n10
Forest Drought Stress Index, 231, 233–35, 237; projected using global circulation model ensemble, 235
forest rangers: early men on Jemez National Forest, 85; problems with hippies, 271–75; *Reminiscences of*, on inner bark use, 294n2, 295n9
forest road: No. 10, 253; No. 376, 84

forests: burned-over areas, 209; resilient, 222; thinning of, 215, 217, 219
Forest Service: acquires San Diego Land grant, 84, 149; acquires Soda Dam, 169, 172–73; acquires SOP lands, 268; Cerro Pelado Fire blame, on, 212; cutting of juniper trees, 223; fatalities of Jemez District employees, 94; managing hot springs and hippies, 272–75; new ranger station in Jemez Springs, 94, 162
fossils: 29–31, 191–95; conifer, photograph of, *31*; in core samples from Banco Bonito explosion pits, 181; crinoid, 191; photograph of brachiopods and gastropods, *190*; Pleistocene-age, 193; in the roof of Spanish Queen Mine, 29; skull, of *Sphenacodon* in Permian red rocks, 31
Franciscan priests: Bartolomé de Ojeda educated by, 18; Fray Alonso de Benavides, 27; Fray Geronimo Zárate Salmerón, 11, 9, 43; at Gíusewa, 9; residence at Patokwa, 22
Fred's Farms, 209
freight wagon: ascending road over Soda Dam, photographs of, *170–71*
French Drain, at Soda Dam, 169, 302n1
Fretwell, Stretch, 13
fuels, forest, accumulated, 205; management of, 199
furnace: arched stone, copper smelting at Spanish Queen Mine, 28, 110

Gardner, Jamie, 163
Giggling Springs, 61
Gila Apache, 99
Gila Wilderness, 99
Gilman Landing, sawmill and teepee burner below Guadalupe Box, 84
Gíusewa, 5, 9, 27, 109; location of San José de los Jémez mission, 127; about Oñate's visit to, 3–5; resettlement of

after 1626 and reconstruction by Fray Martin de Arvide, 284n5; Zárate Salmerón at, 9, 11

Glorieta Pass, Battle of, Civil War, 113

Glover, Vernon, 79; about cost and difficulty of building tunnels in Guadalupe Box, 80; about purchase of timber rights on San Diego grant, 78; his map of SFNW rail line, 76

Goff, Fraser, 301n6, 304n2, 305n6; description and dating of Soda Dam, 300n2; theory of massive travertine formations at Soda Dam, 163; *Travertine Deposits of Soda Dam*, 300n2

Gonzales: Frank, 6–7; family, 93

grant, Cañon de San Diego, 72, 291n2, 296n2; boundary of, 90

grasslands, montane, 211; photograph of, 212

grazing, regulated by Forest Service, 85

greenhouse gases: effects of on climate and droughts, 205, 238; *Impact of Anthropogenic Climate Change on Wildfire across Western US Forests*, 310n1, 307n2; increasing, 237

grizzlies: in the Jemez, 99–105; last ones in New Mexico, 104

grotto: on north side of Soda Dam, 47; on south side of Soda Dam, photograph of, 154

Guadalupe Box, 79–81, 82–84

Guadalupe Mesa, 17–18, 20–25, 109; *Smilodon* fossil found at, 194: Towa name for, 285n1

gypsum deposit, at White Mesa mine, 193–94

Handmaids of the Precious Blood, 131

Hartley Mammoth Site, 193

Hay Canyon, 90

Hayes, Rutherford B., 77

Hemish: people, 2–3, 9, 11, 15, 24, 109, 155, 199, 285n2; 1623 uprising and re-settlement at Gíusewa in 1626, 284n5; ancestral villages, 200, 204, 249; built San José de los Jémez mission, 200; early Spanish contact with, 5; footprint, 2–3; land, forest, and fire use, 203; long-time interest in Soda Dam, 158; placename stories, 317n3; population change over time, 3, 283n1; SFNW railway fruit theft and derailment incident, 83; use of Jemez cave, 175; about Wâavêmâ, 211

Hewett, Edgar, 1910 report on the excavation of San José de los Jémez mission, 284n9

Higgins, Lloyd, and Wanda, and Hummingbird Music Camp, 260

Hillers, John "Jack" K., 33–35; photograph of San José de los Jémez mission, 37; was on Powell's second expedition on Colorado River 1872, 33

hippies: arrival in the Jemez in early 1960s, 269–70; commune at Sulphur Springs, 247; conflicts with locals and Forest Service, *Albuquerque Journal*, "Jemez Hot Spring Is Battlefield: Forest Service, Hippies May Clash Over Nudity," 272–75; dipping with, 276; painted school bus, 271

History Grove, 95–96, 98; name origin, 95

Hog Farm, commune, 269–71; hippies in Spence Spring, Lisa Law photograph, 270; painted school bus in El Rito, Lisa Law photograph, 271

Holmes, William Henry: grizzly bear story, 101; memoirs of, 58; sketch of 1887 campsite, 56

horno (pueblo-style outdoor oven), at *convento* house at old mission ruins, 36, 37

horses: drowning in Jemez River flood, 187; fossil bones of stilt-legged, 192–94; introduction of European species to North America, 194–95; logging with, 145–46, 250; trespass on National Forest, 88

Index

Horseshoe Springs, 96
hotel: 1881 advertisement for Stone, 241; Clay and La Esperanza, 10; first automobile at Stone, 69; Montezuma in Las Vegas, NM, 311n1; Perea manager of Stone, 77; at Sulphur Springs, 52; at Sulphur Springs burned down, 247; Woodgate proprietor of Stone, 69
hot springs: arising at Jemez Fault, 165; enclosed by Otero in Jemez Springs, 244; hippies overstaying at, 271; at Sulphur Springs and National Park Service plans for, 248
Hot Sulphur Water Baths building in Jemez Springs, 61
Hottinger, George, 68
Hughes Mill, lumber, 90
Hummingbird Music Camp, established by Lloyd and Wanda Higgins, 259–60; near John Wesley Powell's 1887 expedition campsite, 37, 58
Huning, Franz: 1857 trip to Jemez hot springs, 110; memoir, *Trader on the Santa Fe Trail*, 297n6
Hurteau, Matthew, op-ed in *Santa Fe New Mexican*, "Use All Tools—Including Fire—to Restore Forests," 219, 309n1
hydrogen sulfide gas, knocked out drillers at Redondo Border well site, 248

Ice Age, animal fossils from, 193
ignimbrite: Battleship Rock made of, 181; from El Cajete vent, 249
increment borer: tree-ring sampling tool, photograph of, 97; used to sample peeled trees, 96
influenza: death of Albert van Derveer in 1918 on Cebolla Creek, 299–300n4; pandemic of 1918, effects on homesteaders on Cebolla Creek, 144

inner bark: uses of as food or medicine, often during hard times, 95–100; isocupressic acid in pine tissues causing abortion in animals, 295–296n9; Lewis and Clark mention of, 95
irrigation: cottonwood trees effects on, 223; on Jemez River dam plan of John Wesley Powell, 57; on Powell's statement to Congress on, 289n3; since 1710 in Jemez Valley, 54
isotopes, unstable, 161

Jemez ancestral villages, 3–5, 283n6, map of, 2
Jemez Canyon Dam, 57
Jemez Cave, 155, 162; child burial found in, 261; damage by climbers, 173; excavation of, 175; photograph of, 164, 177; possible origin of, 176; yucca fiber sandal from, 164
Jemez Country development, 167
Jemez Fault, at Soda Dam, 155, 162, 165
Jemez Forest Reserve, 83, 85, 201–2
Jemez Historic Site, 27, 33–37, 88, 109, 127
Jemez Hot Springs, first crop planted at, 299n3
Jemez Land Company, 292n3, 293n3
Jemez National Forest, 293n1; establishment of and dates, 1907 to 1915, 85; John Adams Sr. mapped boundaries of, 72, 85
Jemez Pueblo, 17, 41, 79, 109–10, 112, 115, 129, 131, 286n8, 290n9, 291n1, 305n1, 305n3; 1882 trip to by Mary Stright, 66; Matachines dancers from, photograph of, 43; railroad right-of-way forced through lands of, 78; retains aboriginal title to Banco Bonito, 71; wildland firefighter fatalities, 94
Jemez Pueblo mission, Presbyterian in Walatowa, 129

Jemez Ranger District: acquired San Diego Grant in 1960s, 75; establishment of in 1920s, 87; fatalities of Forest Service employees, 94; third office built in 2023, 94

Jemez River: 57, 183–84, 224, 278–79; 1941 flood on, 83; beginning at Battleship Rock, 249; effects of cottonwoods on, 223; flood of 1941 washing out of railroad tracks, 305n3; floods of 1941 and 1958, 183; freshet (flood) of 1886, 48; John Wesley Powell's description of, 54; photograph from about 1910 showing, and Soda Dam, *189*; Russian olive trees planted along, 229; view of in 1884 at junction of Rios Guadalupe and Jemez, *227*; view of in 1900 and Soda Dam, *162*

Jemez Springs: 1886 tourist visit to, 45, 289n1; 1903 photograph of Navajo riders in, *110*; 1920 and 1930 census of, 80; 1958 flood isolated, 187; 1960s hippies in, 269; central park and boulevard names of, 268; dangerous place before 1864, 110; first automobile in 1912, 66; Francisco Perea and family move to in 1881, 129; Hemish village in, 3; horse logging above, 145; John Miller moved to, near the old mission ruins, 112; *Large Pueblo Sites Near Jemez Springs, New Mexico*, Elliott report, 320; Los Ojos Bar in, 59; Los Ojos Calientes place name, 109; M.S. Otero visit to in 1903, 246; oldest churches in, 127; Oteros living in after 1882, 131; photograph of forest rangers residence in, *86*; photograph of in 1884, *134*; railroad to, plans begun and dropped, 292n2; ranger district office built in 1929, 87; regarding first settlement of, 297n6; robberies in, 62, 290n8; Rowland's Ranch in 1925, 259; Servants of Paraclete facility established in 1947, 263; Simpson's visit to (Los Ojos calientes) in 1849, 28; stagecoach lines to, 244; Stone Hotel proprietor in 1912, 69; turkeys in, 260; view south in 1884 from a point about the bath house, *226*; visit to by Oñate, 3

Jemez Sulphur mine, 246–47

Jemez Valley: 1689, 1684, and 1696 battles in, 17; oldest photograph in 1879, *33*; Penitentes in, 115; Permian fossils from, 31; Presbyterians move to, 129; railroad in, 77; sheep herds in, 227

Jemez Valley History, web page, 72

Jicarilla Apache, 100, 297n4

Joaquin Mesa, 82–83

juniper, trees, 223, 227

justice, Townshend seeking from Perea, 112–13

Keleher, William, on battles with Navajos, 297n4

Kelly's Ranch, 48, 52

kids: free-range, 16; wild Jemez Springs, 29

killing: of prairie dogs, 141; of priests and soldiers at Patokwa in 1696, 24; of seven priests during 1680 Pueblo Revolt, 19

kiva: pictures of Aztecs on walls at Acoma, 44; legend of sacred fire at Pecos, 289n5; theft of pot from Zia, 38

knowledge and wisdom in stories, 277

Kulisheck, Jeremy, 283n1, 283n3

Kybos, at Camp Shaver, 254; the meaning of the name, 257

La Cueva: flood washed out culverts at, 189; Forest Service guard station at, 87; Vallecitos de, 138

La Junta Fly Camp, 252–53, 257

Index

lakes: above Soda Dam, 155; to be created by six dams planned by John Wesley Powell, 53; in the Valles Caldera, 57, 165–66, 300n2

Lamb, Bruce, 84

Lamy, Jean-Baptiste, archbishop, suppressing Penitentes, 127; transcribed ancient book at Jemez Pueblo, 41

Langley, Samuel P., 56

La Niña, events, 238

Las Conchas: Fire, 205, 207, 211, 214, 307n8; Forest Service guard station location, 87; Powell planned dam location at, 57

Las Trampas Grant, 75

lava flows, 71, 181–82, 279, 285n11; accurate pressure ridges and swales formed by, 178–79

La Malinche, wife or daughter of Montezuma, 39–40

La Ventana, 78

Law, Lisa, photographs of hippies, by, 270–71

Leary, Timothy, 269

Lebanese, Syrian immigrants to New Mexico, 59

Leech, Isabella, 129

legends: of gold, 27; of Hemish transition from Jemez Cave use to caves beneath Soda Dam, 175; of Hemish warriors levitated, 21; of Montezuma in New Mexico, 39

Leopold, Aldo, 229

letters: Archbishop of Santa Fe to Father Gerald Fitzgerald, 315n5; Cartas de Lorenzana of 1776, 41; Fitzgerald to Catholic leadership, 264–65, 314n3; Moses Abousleman to Nathan Salmon regarding his arm, 64, 290n9; Ojeda to de Vargas, 20; Townshend to his wife Dorothea, 132

Lewis: Hazel, 91; Leonard Ward, 89, 91–92, 94–95; Leonard's death reported in a newspaper article, 293n8

Lewis, Meriwether, 95

lidar: 2012 Forest Service acquired data, 304n3; image of contour-plowed forest reforestation, 209; map of Banco Bonito explosion craters, 178, 181; use in forest ecology, 304n4

Liebmann, Matthew J., *Revolt: An Archaeological History of Pueblo Resistance and Revitalization in 17th Century New Mexico*, 17, 283n2; on the ambiguity of which saint intervened at Guadalupe Mesa, 286n9; estimate of corn weight removed from Astialakwa, 287n11; final coalescing of Potokwa and Walatowa, 287n12; regarding mestizo term, 285nn4–5

lightning, ignited wildfires, 200, 203

limestone, marine and terrestrial, 163

Lincoln, Abraham, friend of Francisco Perea, 113

livestock grazing: effect on cottonwoods, 227; effect on fires, 149, 203

Livingston, J. K., 1886 newspaper article by, 45, 174; Soda Dam description by, 52

locomotive boiler explosion on SFNW railway, 81

log chutes, 83; in Area 3, two-thousand-feet-chute, 146–47

logging: 1940s to 1970s on San Diego Cañon Grant, 73–84; begins in 1924 on Rio Guadalupe, 80; camp called Porter, 251; "cut the best and leave the rest," or high grade, 147; dense tree establishment after, 149, 215; by horses and oxen, 145, 250; loggers from southeastern United States, 293n6; trucks, 81, 83–84; view of Rio Guadalupe in 1904 and 2021, 83

logs: stuck in Soda Dam, photograph of, 156–59

Long Walk, Navajo 1864, 297n6

Los Alamos: 205, 211–14; Atomic Energy area, 87; National Laboratory, 163, 300n2

Los Azufres, 244
Los Griegos, 90
Los Hermanos de la Fraternidad Piadosa de Nuestro Padre Jesus Nazareno, 115. *See also* Penitentes
Los Ojos Bar, 59–61, 64–65, 134, 205, 270
Los Ojos Calientes, old name for Jemez Springs, 28, 109
Lourdes, Servants of the Paraclete facility north of Soda Dam, 263
Lucas, Spencer, 29, 194
Lummis, Charles, *118*

Mack truck, loaded with logs, *81*
Madalena, Joshua, story of infiltrator-spies at Battle of Astialakwa, 286n8
Madera limestone, 163, 190–91
magma chamber, below the Valles Caldera, 166
maize agriculture, earliest known in northern NM, 155
mammoths in the Jemez, 193–94
map: of 1906 farmland parcels in San Diego Grant, 291n6; of Astialakwa and Patokwa on Guadalupe Mesa, *21*; of Cañon de San Diego Grant plat, *70*; of Hemish Ancestral villages, *2*; lidar map of Banco Bonito, *179*; new geologic, of Valles Caldera, 301n6; of precipitation anomalies in United States, 2015–2023, *230*
marijuana, 255, 269
Martinez, José Antonio Padre, 127–29, 299n4
master tree-ring chronology, 96–97, 203
Matachines, Jemez Pueblo dancers, photograph of in Jemez Springs, *43*
McGlothin, Charles, 172
McMillen, Alonzo B., lawyer in New Mexico land grant takings, 73–75, 293n3

medicinal: springs, 243; treatment for priests, 266; uses of inner bark, 96, 100, 243, 295n9
Medio Dia Canyon and Creek, 104; photograph of log chute in, *146*
megadrought, 220, 235, 237
megafloods, 184
melon thieves at Jemez Pueblo, 83
memoirs: Franz Huning, *Trader on the Santa Fe Trail*, 110; William Henry Holmes, *Random Records*, 58, 289n4
memory palace technique, 277–79; method loci, 317n4
Menaul: Historical Library, 299n7; School, 127
Mera, Reba, 37
Merhege, John, 311n10
mesa: Borrego, 283n6; Cerro Colorado, 19; forest thinning on, 219; Guadalupe, 20, 24; gypsum mine on White, 193; logging on, 147; plowed contours on Cebollita, south of La Cueva, 209; Virgin, in Holmes sketch, *56*
Mesa Verde, 88
mess hall, at Camp Shaver, *253–56*
mestizo, 17, 285n4; woman, 39
Mexico City, 9, 39–42
military road to Valles Caldera, 244
Miller: family, photograph of, *130*; Hugh, 10, 130, 144; John, at Jemez Pueblo, 112, 119; John, moved to Jemez Springs, 131; John, Hugh, and Townshend buggy ride to Rio Cebolla, 1903, 137–39; John, photograph of at Fentons, *139*; John, photograph of at Fentons with boys, *144*; John, prairie dog killing and eating, 141; John, "won't allow no buck to bite him," 141; Mary (née Stright), 10, 130; Townshend visits in 1903, 114
minerals: Alonzo de Benavides statement about, in Jemez, 27; claim, with sulphur, 244

Index

mines: Burnett Company report, Spanish Queen, 28; sulphur, 243–47
mission, San José de los Jémez, 9; artist's rendition of in 1620s, 8; at Gíusewa, 5; map and diagram of, 4; *National Historic Landmark* document, 283n1
mnemonic. *See* memory palace
Momaday, N. Scott, allegorical Correr El Gallo story in *House Made of Dawn*, 298n4
Montezuma: El Monarca, 39; legends and myths of in New Mexico, 39–44, 288n1, 288n3, 289n4, 289n6–7; flies from New Mexico to Mexico City on an eagle, 41–42; myth as Mexican government propaganda ploy, 40
Montoya, Mariano Sheriff, 64–65
Monument Cañon, 49–52, 283n5, 289n1, 303n1; nearly forgotten place of the Jemez, 52; Padre Alonzo Trail to, 5; repeat photograph set, *50*; visit to by Livingston family, 45
Moore, Earl, 87
Morgan, Gary, 193–94
mule: logging with, 146; William Henry Holmes story of, and a grizzly bear, 102
mummified child burial in Jemez Cave, 261
murder: of Archuleta sons by Navajo, 136; of Mangus Coloradas at Piños Altos by US soldiers, 99; of Placitas commune members by Donald Waskey, 271, 317n2; of Richard Wetherill Sr. at Chaco Canyon by a Navajo, 88; of Vicente Elias by political foes in Cañon, 112, 114, 116; of young woman by Father John B. Feit in Texas, 267, 316n11
mysteries: Alice Bullock book about New Mexico, 295n9; cause of 2022 Cerro Pelado Fire, 212; existence of modern Soda Dam, 165; use of inner bark, 96, 100

Nacimiento: Mountains (or Range), 233, 277; village of, 136
Nanishagi (Nona Shagi), 252, 313n7
National Park Service, 213–14, 248, 268
National Science Foundation, 199
National Youth Administration, 253; NYA/UNM camp at Battleship Rock, 254, 257
Navaho Expedition of 1849, 108, 288n3
Navajo, 19, 88, 99, 110, 297n6; Archuleta sons killed by, 136; family dwelt in the Jemez, 23; fear of by Franz Huning, 112; historical accounts of raiding by, 15, 297n4; photograph of riders in Jemez Springs, *110*; raids before 1650, 127
Nee Hemish, history by Joe S. Sando, 17
New Deal, programs, 8, 252–53
New Mexico: 1870 Territorial Government law allowing partition of community land grants, 72; 1879 BAE expedition to, 33; 1885 *Attractions of*, book, 41; 2022 wildfires in, 219; AT&SF arrival in Las Vegas in 1879, 77, 243; Correr El Gallo in, 121; de Vargas Army in, 19; dinosaur fossils in, 191; Donald Waskey/Ulysses Grant runs for Governor of, 270; drought and flood history in, 53, 233; forest restoration in, 308n11; last grizzly bears in, 105; *Los Arabes* of, 290n1; mixed heritage people in, 285n4; Montezuma legends in, 39; Padre Martinez controversial in, 129; pedophile priests in, 266, 316n7; Penitente history in, 115; place names of, 283n7; Presbyterians in, 127; reforestation initiatives in, 210; Santa Fe Ring in, 72; Senator Holm O. Bursum, 78; *The Tenderfoot* in, 114; University of, 5, 173, 259; UNM archaeology, field camp photograph of, *250–51*

Index

New Mexico (Highway) Department of Transportation, 167, 302n1
New Mexico Land & Timber company (NML&T), 81, 83–84, 292n3
New Mexico Magazine, 37
New Mexico Reforestation Center, 210
New Mexico Wildlife Federation, 172
New Spain, 11
Norteños, 211
Nusbaum, Jesse: photograph of San José de los Jémez mission, *10*; photograph of Soda Dam, *47*; photograph of Soda Dam with hanging log, *156*
NYA. *See* National Youth Administration

oats: domestic, 55; wild, 138
Ojo de Chihuahua, 5
orchard, Rowland's, north of Soda Dam, 259–60
Otero: bathhouse, 134; Dolores, 113; Mariano S., 72–73, 77, 129, 189, 243–47, 292n2, 310n4, 311n2; Miguel A., 292n2; Miguel A. II, Governor, 85; and Francisco Perea family, 131; builder-owner of Stone Hotel, 134

Padre Alonzo Trail, 279; leading to Monument Canyon, 52; Oñate travels on, 5
Pajarito Plateau, 87, 235, 237, 244
paleo-canyon, formed Battleship Rock, 249
paleontologists, paleontology, 29, 191, 193–94, 306n2
Paliza CCC camp, 253
parishes: abusive priests returned to, 263–64, 266–67, 315n6; in Jemez Springs, 267; in New Mexico, 266
Park, Fitzgerald, in center of Jemez Springs Village, 317n13

partition of community land grants, 72, 75
Patokwa, 18, 21, 287n12; Hemish return to after 1696 revolt, 24; Hemish return to after Battle of Astialakwa and captivity, 22; reoccupied after 1680 Pueblo Revolt, 20
Pecos, Anthony, 94
peeled ponderosa pine trees, 96, *97*, 99–100, 294nn1–9
Pejunkwa, 283n6
pelado, bald mountains, 211, 215
Peña Blanca, 244
Penitente: ceremony, 115–119, *118*, 129; Chavez, "The Penitentes of New Mexico," 298n1
Perea: Francisco, 48, 77, 112, 129, 131–34, 243–44, 297n11; Dolores Otero, first wife, 113; Gabriela Montoya, second wife, 134; José Leandro Perea Sr., his uncle, 129, 133; moved to Jemez Springs in 1881, 129; New Mexico territorial delegate, 244; photographs of, *132*; photograph of home, *134*; "Pereas have all lost their money somehow," 133; José Ynez, brother of, first ordained Hispanic Presbyterian minister in New Mexico, 129, 299n7; José Ynez, establishes Jemez Pueblo mission, 129; José Ynez, present at Sandoval baptism, 131
petroglyphs, 24–25; Dancing Coyotes, 287n12
Pettit, Roland, story of underwater passage beneath Soda Dam, 176
Picket Wire (Purgatory River), Colorado, 140
Picuris Pueblo, 11, 100
pile burns, 212; holdover fire, 212, 219
Pinchot, Gifford, 85
piñons, 227; scrubby, 47
placename stories, 317n3
Plana, Miguel and Dee, 189, 245

343

Pleistocene fossils, 194
ponderosa pine, 6, 95, 100, 198, 209, 327; forests, 71, 200–201; peeled, 95–97, 100, 295n9; plantation of, 209; rejected, photograph of, 148
Porter, James, 266
Porter Landing, logging camp on Rio Guadalupe, 251
postal service, Star Route to Jemez Springs by 1894, 66
post-fire erosion, photograph of, 220
pot, missing Zia sun, 36, 38
Powars, W.F., Colonel, 103
Powell, John Wesley, 1887 expedition to the Jemez, and campsite sketch, 56–58; 1869 Colorado River run, 53; appointed USGS director in 1881, 57; grandiose Jemez dam plan, 54–57; horseback trip to Redondo Peak, 58; ideas about dams and water storage in the western US, 53; oral testimony to Congress, 54; photograph of, in Flagstaff 1891, 55
prairie dogs: innumerable in La Cueva in 1903, 138; killing and eating, 141
Precambrian granite, at Soda Dam, 169
Presbyterian church: 1884 photograph of, 134, 298n2; elders, 133; history in New Mexico and Jemez, 127, 299n2; Jemez Springs Community, records of, 299nn6–7, 299n9; mission schools, 127, 129; missionaries, 127, 131; Reverend Elijah Maclean Fenton presiding at, 131; Reverend John Shields establishing in Jemez Pueblo and Jemez Springs, in 1878 and 1881, 129
prescribed fires, 149, 214; escaped, 219
priests: abusive, 263–67, 314nn2–3; few in New Mexico after 1821, 115; killing of at Potokwa in 1696, 24; killing of seven during Pueblo Revolt of 1680, 19
psychedelic-painted school buses, 269, 271
Pueblo Bonito, 88

Pueblo Lands Condemnation Act of 1926 (and 1928), 78, 293n4
Pueblo Revolt, of 1680, 17–18, 109, 194, 285n2; uprising of 1623, 284n5; Weber, *What Caused the Pueblo Revolt?*, 285n4
pumice, 181, 194, 301n6; mining, 201
Puname warrior allies, Zia, Santa Ana, and San Felipe, 20–21
pyroclastic, fall and flow deposits, 181

radiometric dating: methods, 161; of Valles Caldera volcanic events, 304n1
railroad: logging, 81, 149; SFNF, 74, 77–81, 83, 243–44, 293n6
ranch: Abousleman, 62; R.B. Townshend's in Colorado, 112; Rowland, 260; Sandoval, 139; a staple "ranch" joke, 141
rangers: diaries of early southwestern forest, 293n7; "forest arrangers," as called by Gifford Pinchot, 85; first office (station) built in Jemez Springs in 1929, 87–88; Fred Swetnam, 92; Leonard Ward Lewis, 89; ranger's day in the office sign, 92; new station built in 2023, 94; Perl Charles first on Jemez River District, 87; residence in Jemez Springs in 1929, 86; Rodriquez on Coyote District, 89; sometimes they packed side arms, 85; summer with families at Seven Springs guard station, 93
Raynolds, Joshua S., Albuquerque banker in land grant takings, 74–75, 293n3
Read, Benjamin, 41
reconquest: battles of, in 1689, 1694, and 1696, 17, 19; failed attempt in 1689, 18; Ojeda's role in, 19, 286n6; Spanish-Pueblo allies during, 21
Redondo Border, 248

Redondo Campground, 179; map of explosion craters near, *181*
Redondo Creek, 292n2, 327
Redondo Peak, 279; 1876 USGS map name of Pelado, 211; 1903 photograph from Area 3, *150*; Hemish eagle image on, 211; Hemish name for is Wâavêmâ, "Want for Nothing Peak," 211, 279, 308n3; Holmes and Powell horseback ride up in 1887, 58; photograph of Thurber fescue grassland on, *215*
reforestation: as carbon-capture, 239; following severe wildfires, 209
Reiter, Paul, excavation of Jemez Cave by, *164*, *177*, 300–301n3, 314n5, 314n7
Relaciones, by Fr. Zárate Geronimo Salmerón, 11, 43–44
reversed topography at Battleship Rock, 249
Rinehart, Larry, 194
Rio Cebolla (Cebolla Creek), visit to, by Millers and Townshend, 57, 131, 137–44, 251, 299n3
Rio Chama, 278
Rio de Las Vacas, 87, 251
Rio Grande: Google Earth image showing, *278*; oldest historical floods on, *183*; wildfires in the bosque, 223
Rio Guadalupe: logging along, 78; view just below the Box in 1944, *228*; view of the junction with Rio Jemez in 1884, *224*; view up in 1904, *83*
Rio Jemez. *See* Jemez River
Rio San Antonio. *See* San Antonio Creek
Ritch, William G., 41–42, 289n5
road: ascending west end of Soda Dam in 1910, *47*; dirt, below Soda Dam in 1920s, *189*; logging, on benches above Jemez Springs, *147*; view of, over west end of Soda Dam in 1903, *171*; washed-out, *187*

robbery, of Abousleman stores, 64; of Abousleman sheep, 62
rock climbers, in Jemez Cave, 173
Rodriguez, Joe, 89
Rogers, Rosa, 255
Romer, Alfred Sherwood, fossil discovery in Spanish Queen Mine, 29
Romero: Barnabe, 287n13; family, 93
Romney, Hugh, 269. *See also* Wavy Gravy
Roos, Christopher I., 200, 306n2, 307nn6–7, 307n10; core sampling explosion crater on Banco Bonito for fire, forest, and human history, *180*, *181*; *Native American Fire Management at an Ancient Wildland–Urban Interface in the Southwest United States*, 305n8
Roosevelt: Eleanor, 253; Theodore, 77, 85
rooster, running of the, 122, 124–25. *See also* Correr El Gallo
Rossetti, Stephen J., accounting the number of priests treated for child sex abuse, 315n6
Routledge: Joesph E., 259, 317n4; grizzly bear story, 104; quote about springs at Soda Dam, 303n3
Rowland, Howard, estate value, 314n4; family, 314n1; guests and turkeys, 259–60
Russian olive trees planted along Jemez River by Boy Scouts, 229

saber-tooth cat (*Smilodon gracilis*) fossil, 194
sacred: fire, 289n5; Hemish shrine, 58; Zia pot, 37
Saint Didacus (San Diego), 286n9
Saint Joseph, 11, 15. *See also* San José de los Jémez mission church
Salmerón, Geronimo Zárate, 44; construction of San José de los Jémez

Salmerón, Geronimo Zárate, (*continued*) mission church, 9; left the Jemez in 1626, 11; mention of Aztecs visiting Zia, Jemez, and Acoma Pueblos, 43; report on mineral deposits in the Jemez, 27; survived the 1623 uprising, 284n5

Salmon, Nathan, 61–62, 64, 290n9; partnership with Abousleman, 59

San Antonio Hot Springs, 275–76

San Antonio Creek, 179, 249; hippies camping on, 269

San Antonio Mountain: lava flow from, 165; lava flow from damming the East Fork, 249

Sanchez, Frederick R., summarizes Presbyterian church history in New Mexico, 299n2

sand dunes, in lower Jemez Valley, 54

Sandia Mountains, 236

San Diego: apparition of saint (or Santiago), on Guadalupe Mesa, 25, 286n9; church of, 11; del Montes, 22; de los Jémez at Walatowa, 24; saint, levitating falling warriors in 1694, 21

San Diego Canyon (or Cañon), 3, 4, 9, 18, 77, 109–10, 111, 155, 223–25, 226, 250; Simpson travels up in 1849, 28, 108, 109

San Diego Grant (or Cañon de), 70–78, 88, 109, 268; Forest Service acquires, 84; land grantees, 297n6; land grant heirs, 74–75; partition and sale, 74, 291n6; sawmill on, at Battleship Rock, 312n4; sheep grazed on, 310n4; timber and logging on, 78, 147, 250–53; timber sale values on, 293n3

Sando, Joe S., on burros "braying like coyotes" in 1694, 287n12; on SFNW railroad land grab, 293n4; on 1941 flood impacts, 305n1; on SFNW railroad land grab, 78; on Zárate Geronimo Salmerón, 11; on Pueblo Revolt history, 17; story about train engineers stealing watermelons, 83

Sandoval: Francisco, 28; Genevieve, 187; Gilbert, 93, 187; Juan, 137–39, 299n3; Juan and Juliana, baptized Presbyterian in 1880, Reverends John Shields and José Ynez Perea presiding, 131; Juanita, 93; Moises, 93–94; Simon, 93; stalwart sons of Juan, 133

Sandoval County, 68, 194; sheriffs, 61, 64, 212–13

Sangre de Cristo Mountains, 6, 75, 77, 212–13

San Felipe Pueblo, 18, 20

San José de los Jémez mission church, 5, 9, 11, 43, 127, 130; artist rendition of during 1620s, 8; ceiling of plaited willow work, 285n11; diagram of ruins. on 1875 map, 4; *National Historic Landmark* nomination, 15, 283n1, 285n10; photograph by J. K. Hillers 1879, 34; possibly abandoned after 1632, 15; restoration of parts of in 1936 by CCC, 253, 284n9

San Juan: feast day, June 24, 122; Oñate's real, at the Pueblo of, 5

San Juan Mesa, 96, 149, 212

San Mateo: Mountains, 105; village of, 118

San Pedro Parks, 277

Santa Ana Pueblo, 18, 20, 57

Santa Clara Ranger District, 87

Santa Fe: child burial from Jemez Cave displayed publicly in, 261; de Vargas re-occupied 1693, 19; Livingston family trip to Jemez from, 1886, 45; mining equipment hauled to Sulphur Springs from, 244; Otero political and business insider in, 77; ox carts brought to Potokwa to haul corn from Astialakwa, 22; Salmon-Abousleman store in, 59; Zia sun pot turned up in, 38

Santa Fe National Forest, 72, 304n3, 318, 321, 327; Jemez National Forest combined with, 85; Jemez River Ranger District established 1929, 87; logging

on, 84; San Diego Cañon grant added to Jemez Ranger District, 75
Santa Fe New Mexican newspaper: "Early Alarm about Church Abusers," 2022, 314n2; fire in Abousleman home and store in 1900, 290n7; gun battle between Moses Abousleman and sheep thieves 1912, 61; Livingston family trip to Jemez 1886, 45; New Mexico developing Reforestation Center 2023, 308n12; op-ed by Hurteau, Swetnam, and Allen about fire management, 309n1; "Runner Attacked by Bear in Valles Caldera" in 2016, 296n1; Salmon-Abousleman advertisement 1900, 60; "Two Dead in Midnight Duel" in 1912, 290n9
Santa Fe Northwestern Railroad (SFNW), 78; bought by New Mexico Lumber and Timber in 1930s, 81; incorporated in 1920, 78; *Jemez Mountains Railroad* report by Glover, 292n2; Joe Sando regarding right-of-way issue through Jemez Pueblo lands, 293n4; map of the rail lines, from Glover, 76; railroad tracks, torn out by 1941 flood, 228; spectacular and expensive construction of tunnels, 80
Santa Fe Ring, 72, 74
Santiago: photograph of, 25; the saint (or San Diego), 21, 286n9; white beard astride, flowing, 286n9
Santo Domingo Pueblo, 7
San Ysidro, 28, 66–67, 78, 193; fight with sheep thieves near, 62; Jemez River at, 54
sawmill: at Battleship Rock, 250, 257; at Gilman landing, 84; just above Soda Dam, 48
Scanland, J. M., *Theft of the Sacred Fire*, 289n5
scars on ponderosa pines: fire-caused, photograph of, 198; peeling by people, photograph of, 97; Christian crosses carved by sheepherders, photograph of, 98
Scurlock, Dan: floods on the middle and upper Rio Grande, 183; on James Smith's sawmill at Battleship Rock, 312n4; numbers of livestock on Baca Location in 1918, 310n4; regarding 1941 flood, 184; regarding an aborted plan to build a rail line to Jemez Springs, 292n2
Seligman, Carl, 172
Servants of the Paraclete: Church Canyon property, 5; documentary sources on sexual abuse history, 315n5, 316n7, 316n9, 316n12; donation of land for Jemez Springs village park, 317n13; Edwin Byrne, co-founder of, 264; Fitzgerald elbowed out of leadership in 1965, 266; location of main buildings, 10; newspaper articles about, 314n2; old hospital, now National Park Service building, 268; recycling of pedophilic priests back into parishes, 267; recidivism by treated and recycled priests, 266, 315n6; scandals of, 263–68
SFNW. *See* Santa Fe Northwestern Railroad
Shaver, William, 254
sheep: number, grazed on Baca Location in 1918, 310n4; grazing on San Diego land grant, 73, 244; grazing on cottonwood seedlings, 227; herding by Moses Abousleman, 59; hundreds seized by Spanish after Battle of Astialkwa in 1694, 287n11; large bands of, 59, 227; massive loss of herds due to April snowstorm, 134; more than 1,000 in San Ysidro in 1882, 67; stolen by Navajo, 297n4; thieves of, 61–64
sheepherder cross carvings on ponderosa pine trees, 98

347

Index

Sherman, William Tecumseh, 77
Shevenell, Lisa, 300n2, 301n1
Shields: family history in Presbyterian church records, 298–99n2; Helen, married John Amos Adams, 131; Helen Belle, 87, 131; Isabelle, 66; John Milton, 66, 72, 87, 129; Linus, 312n6; photograph of family, 128. *See also* Adams family
shrublands, conversion of forests to, 213, 307n10
siege, at the Battle of Astialakwa 1694, 20. *See also* Astialakwa
Sierra de los Pinos, 5, 149, 211, 217, 219–20; Google Earth image of, 218
Sierra de los Valles: map name in 1876 Wheeler Survey, 317n1; paradoxical but lyrical place name, 277; sulphur deposits to the west of, 244
silver: deposits in the Jemez, 27; Spanish Queen Mine production in 1928 and 1929, 28
Simpson, James, 28, 108–9, 288n3–4
sketch: 1875 map of San Diego Cañon and of San José de los Jémez, 4; of Powell-Stevenson expedition campsite, 56
slash piles, Cerro Pelado Fire, 219
Smith: Clyde, 249–50, 310n4, 312n4; James, 134, 249, 312n4
Smithsonian Institution, Washington, DC, 33, 56, 270–71
Soda Dam, 158, 302n1; 1886 description of, 45; 1910 Nusbaum photograph of, 47; blasting of, 167; caves of, 174; freight wagon ascending road in 1903, 168, 170; geology of, 153; grizzly bear sleeping in cave of, 103; hot springs on and around in 1886, 50; logs and floods, 155; master plan for, 172–73; modern, 155, 161–63, 166, 176; north side of, 159, 174–75; photograph of log jam from flood, 185; protecting, 158; road blasting construction records, 302n1; story of dry cave under, 176–77; theory of formation, and lakes in the Valles Caldera, 161–65, 176, 181, 300n2; top of, 164, 165, 168; view of older deposits, 160, 162
soils, formed by ancient grasslands, 213
song, official New Mexico State, 39
SOP. *See* Servants of the Paraclete
Southern Jemez Plateau: forests and people on, 199; Hemish diaspora from, 24; map of Hemish ancestral villages on, 2; population estimate pre-Spanish, 200
South Mountain: volcanic lava flows from blocking East Fork, 165, 179, 249; view of in 1903 from Los Griegos, and in 2023, 201
Southwest: 1897 Stevenson-Hillers expedition to, 33; drought since 2015, 231; ENSO effects on climate in, 238; ENSO-wildfire correlations in, 238; forest conversion to shrublands and grasslands in, 307n10; FDSI from 1000 to 2007 CE graph, 234; graph of rainfall history in, 230; horses brought by Spanish to, 195; last grizzlies in New Mexico, 105; Montezuma legends in, 39; recent drought driven by warming temperatures, 237
Spain, 39, 72, 115, 119
Spanish: alcaldes and governors, 15; ancient diggings at Sulphur Springs, 246; Entrada, 3, 40; Influenza of 1918, 144; land grant heirs, 75; and Mexican courts and laws regarding community land grants, 72; pueblos allied with, 22; reoccupation after 1692, 286n6; soldiers on Guadalupe Mesa, sketch of, 22; twenty families listed on Cañon de San Diego grant document, 109

Spanish Queen Mine, 27–29, 109; fossils in, 31, 191; stone walls, photograph of, 26–27
species: native, 209; non-native plants, 308n11, 327; tamarisk, 309n3
Spence Springs, 276; "Clothing Optional" sign, 276; Forest Service designated nude or bathing days of the week in, 272, 274; hippies sitting in 1967 photograph, 270
Sphenacodon ferocior, fossil of, 29, 31
stagecoach, line to Jemez Springs, 59, 61, 244, 246
staircase, spiral, in tower of San José de los Jémez mission church, 13, 16
state flag design, New Mexico, 36, 37
Stephenson, Catherine: Dancing Coyotes petroglyph, 287n12; on San Antonio Springs, 276; on Soda Dam preservation, 172; story of ticketing nude hippies at Spence Springs, 272
Stevenson: James, 33, 56, 58, 288n2; Matilda Coxe "Tilly," 33, 35, 37–38, 56, 288n4, 288n6
Stone Hotel: 1881 newspaper advertisement for, 242; construction of in 1881, 77; Francisco Perea assumes management of in 1881, 129; loses west wing in 1958 flood, 189; Mr. Woodgate proprietor of in 1912, 69; now part of Bohdi Manda Zen compound, 114; tree-ring dating of floor joists, 311n5; view of from old mission, 13
Stright, Mary (Miller), 1882 buggy trip from Bernalillo to Jemez, 66–67; with husband John and son Hugh Miller, 130, 131
sulphur, 48; about the mine, 243–44, 279; Los Azufres, 244; Manuel Abrego living at in 1857, 112; nuggets and crystals described by Livingston in 1886, 49
Sulphur Springs, 52, 244–47, 249, 292n2, 311n8

Sutherland, Steve, 83, 228
Swetnam: Fred, 88, 92, 93–94, 205–6; managing hippie problems, 272–75; Grace, 92, 294n9; Jim, 92, 254, photograph of at Camp Shaver, 255; Tyson L., lidar study of Jemez forests, 304n4
Syrian immigrants to New Mexico, 59

Taos: Padre José Antonio Martinez of, 129; Hog Farm commune at, 269–70; Pueblo, 200
Tartaglia, Leonard, 172
temperature and precipitation data from the Jemez, graphs of, 233
Tenochtitlan, 39–40, 44
Tenth Circuit Court ruling on Jemez Pueblo land claims in 2023, 71
terraces: moisture-retaining, 209; plowed, for reforestation, 210
Theft of the Sacred Fire, tall-tale about the Jemez, 289n5
thinning, of small diameter trees, 149, 204, 215, 219–21
Thomas, A. E. "Tommy," 172
Thornton, New Mexico, 244
Thurber fescue, grassland, photograph of, 215
time immemorial, Hemish ancestors present in Jemez, 155
Toledo, Frankie, 94
Tosa, Paul, 285n1, 308n3; description of placenames, 317n3
tourists: plan to attract them to Jemez Hot Springs in 1880s, 243; Sulphur Springs a destination for, 52; stage line to the hot springs, 59
Towa, name for Guadalupe Mesa, 17–18, 285n1; name for Redondo Peak, 211
Townshend, Richard B.: books by, 114; Victorian-era prejudices and predilections, 132, 137

trail: 1903 up San Antonio Creek, now State Road 4, 48; ancient paths near Battleship Rock, 312n2; Padre Alonzo, 5, 52; up south side of Guadalupe Mesa, 20, 24; up to Virgin Mesa from Jemez Springs, 279; to Valle Grande, via Padre Alonzo, on 1875 map, 4

Trans-Jemez Road, built by CCC, 253

travertine: basin in Soda Dam grotto, 174; cascade domes over Jemez River, 158; dam formation episodes, 300n2; dating of different dams formed in San Diego Canyon, 161; deposits, 155–56, 165; floods washing out and underneath, 176; growth rate estimate of Soda Dam dome over the river, 301n4; logs stuck on upstream side of Soda Dam photograph, 159; primarily formed of calcium carbonate, 163

tree-ring crossdating, 96–97, 198, 203

tree-ring dates: of beams from San José de los Jémez mission church, 200; of a Christian cross in History Grove, 98; of fire scars on pine tree, 198; of peeled trees in New Mexico, 100

trees: ancient, in History Grove, 95; invading in montane grasslands, 214

trestle, below Guadalupe Box, 79

truck logging operations by NML&T on San Diego Grant and Baca Location, 80–83

Trujillo, Juan, 136

tug-of-war, in Correr El Gallo, 122–24, 298n4

tunnels: in Guadalupe Box, 79, 81; in Spanish Queen Mine, 28–29

turkey: feather blankets made by Puebloans, 261; feathers tied with yucca cordage in Hemish style, 258; white domestic, escaped from Rowland's Ranch, 260; wild Merriam's, 260–61

type conversion, of forests to shrublands following wildfire, 209, 307n10

unintended consequences, of using non-native species, 308n11

Union Army, 129; Francisco Perea, Lt. Colonel, Civil War, 113; James Stevenson, 2nd Lieutenant Civil War, 37; John Wesley Powell, Major, artillery officer, Civil War, 53

Union Oil, geothermal drilling in the Jemez, 311n10

United States Geological Survey (USGS), 37, 213, 219, 305n1; John Wesley Powell, second director of, 53

University of New Mexico: archaeological field school (camp) at Battleship Rock, 251–53, 313n7; landscape architecture class creates plans for Soda Dam, 173; Lansing Bloom, professor of history at, 5

Unshagi, ancestral Hemish village, 252, 312n2, 313n7

Ursus: americanus (black bear), 101; *arctos horribilis* (grizzly bear), 101; *arctos horribilis nelsoni* (Mexican grizzly bear), 58

US Army: defeat of Navajo in 1864, 110; dragoons defeated at 1854 Battle of Cieneguilla, 100; pursuit of Mangus Coloradas in 1862, 99; Wheeler Survey, 211

US Census, 1920 and 1930 in Jemez, 80

USDA Soil Conservation Service, 229

US Forest Service: acquired San Diego land grant, 75, 84; creation of by Gifford Pinchot and Theodore Roosevelt, 85; John Adams Sr., surveyor with, 131; purchases land with Padre Alonzo Trail, 5

US General Land Office, 70

Uto-Aztecan languages, 43

Vallecitos: Creek, 253; de los Indios (Sierra de los Pinos), 5, 52, 201; Viejo

(Ponderosa, New Mexico), 5–7, 283n6
Valle Grande: CCC camps near, 253; Juan de Oñate travels through in 1598, 5; plan for dams and lakes in, 57; view into from Redondo Peak, 215
Valle San Antonio, 57
Valles Caldera: geothermal reservoir in, 163; ancient lakes in, and Soda Dam, 161; peaks in, 211; radiometric ages of volcanism in, 304n1; sheep grazing in, 310n4; theory of lake stands and Soda Dam formation, 300n2; as a vast water storage tank, envisioned by J. W. Powell, 53
Valles Caldera National Preserve: Banco Bonito aboriginal title, 71; bear attack of a runner in, 101; Board of Trustees, original, 98; History Grove, 95; Mariano Otero purchased entire Baca Location by 1899, 129; plans to develop Sulphur Springs for visitation, 248, 311n11
van Derveer, Albert Phillip, 144, 299n4, 300n4–5
vapor pressure deficit, 233
Venado Fire, example of the efficacy of fuels treatments reducing fire severity, 219
Victorio, Gila Apache leader, 99
Virgin Mary, 15
Virgin Mesa, 20; Amoxiumqua, Hemish ancestral village on, repeat photograph set, *200*; Cebollita Fire of 1971 on, 205; Cruiser Rock from Camp Shaver photograph, *256*; W. H. Holmes grizzly bear encounter below, in 1887, 101; horse logging on benches below, 145; stunted, reject ponderosa pine on, *148*; tent rocks below, in drawing, *107*
Virgin of Guadalupe, on east-facing cliff of Guadalupe Mesa, 24, 286n9

volcanic eruptions, forming Banco Bonito, 165
VPD (vapor pressure deficit), 231, 234, 238

Wâavêmâ: Hemish Towa name for Redondo Peak, 211, 279; meaning in Towa explained by Paul Tosa, 308n3, 317n3. *See also* Redondo Peak
wagons: caught in flood of Jemez River, 187; loads of fuelwood, 227; of mining equipment hauled to Sulphur Springs from Santa Fe, 244; what horse-drawn are capable of, 138
Walatowa, 122; became the principal pueblo after the departure of Fray Martin de Arvide, 11; Cerro Colorado, defensive mesa and Zia refugee pueblo to west of, 19; coalescing of Patokwa and Walatowa in the mid-1700s, 24, 287n12; Correr El Gallo at, 121; Hemish return to from Hopi in 1716, 22; recent population of, 2; SFNW railroad tracks built through farmlands of, 78; State NM 4 built along eastern edge of, 84
Walatowan, newsletter story of Battle of Astialakwa by Paul Tosa, 285n1
Walchia piniformis, fossil conifer from Spanish Queen Mine, 31
Waquie: Andrew V., 94; Benjamin P., 94; family, 93
warming, climate, 204, 215, 221, 231, 237–38
Warm Springs: McCauley (Abousleman) near the East Fork, 269; near Picuris Pueblo, 295n9
warriors, at Battle of Astialakwa, 21
Waskey, Donald, 270–71; also known as Ulysses S. Grant, hippie commune leader, 270, 274
waterfall: northwest side of Battleship Rock, 312n5; at Soda Dam, 156, 158

Index

water flumes from dugout logs, 6. *See also* canoa

watermelon theft by SFNW railroad men, 83

waters, meteoritic and hydrothermal, 165

Wavy Gravy (Hugh Romney), at Woodstock, and Hog Farm commune leader, 269; his painted school bus, 271

Westerling, Anthony LeRoy, *Warming and Earlier Spring Increase Western US Forest Wildfire Activity*, 307n2, 310n1

Western Apache, 277

west mesa of Albuquerque, 54

Wetherill, Richard Sr., and Jr., 88

Wheeler Expedition (or Survey), 4, 50; and surveyors, 52

wheezy harmonium in Presbyterian church, 133, 279

whirlwinds, wildfire, 207

White, Leslie A., 43

Whitely, David, former pastor Jemez Springs Community Presbyterian Church, 298n2

White Mesa mine, 192–94

White Pine Lumber Company, 74, 78, 251, 292n3, 312n4

Whitman, Walt, "Weapon, shapely, naked, wan, from earth's midmost bowels drawn," 146

wildfire: accidents, 94; extreme behaviors, 207; large, 205, 307n2; problems, 221; risk, 204, 222; runaway, 205

Williams, A. Park, 231, 233–235, 237, 239; "Temperature as a Potent Driver," 310n2

willows: along Cebolla Creek, 140; along the Jemez River today photograph of, *186*; grown up along the Jemez River photograph of, *83*, *189*

wisdom: Keith Basso, *Wisdom Sits in Places*, 277

Wittick, Ben, 154

Wolfe, Thomas, "You can't go home again," 132

Woodgate, John, 69

Yellowstone of the Southwest (the Jemez), 243

YMCA's Camp Shaver, 249

Yucca: cordage, photograph of with turkey feathers, *258*; fiber sandal, *164*

Zia people, 17–20, 62, 290n8

Zia Pueblo, 37; 1689 battle near, 18; 1918 battle with sheep thieves near, 61; Ice Age animals lived near, 193; Leslie A. White worked at in 1940s, 43; Matilda Coxe Stevenson visit in 1890, 38; sun symbol, New Mexico flag, and missing pot, *36*, 38

Zuni Pueblo, 11, 37, 99